Fail-safe

By Adrian Demaid

This publication forms part of an Open University course T207 *Engineering: mechanics, materials, design*. Details of this and other Open University courses can be obtained from the Student Registration and Enquiry Service, The Open University, PO Box 197, Milton Keynes MK7 6BJ, United Kingdom: tel. +44 (0)845 300 60 90, email general-enquiries@open.ac.uk

Alternatively, you may visit the Open University website at http://www.open.ac.uk where you can learn more about the wide range of courses and packs offered at all levels by The Open University.

To purchase a selection of Open University course materials visit http://www.ouw.co.uk, or contact Open University Worldwide, Walton Hall, Milton Keynes MK7 6AA, United Kingdom for a brochure: tel. +44 (0)1908 858793; fax +44 (0)1908 858787; email ouw-customer-services@open.ac.uk

Fracture Training Associates in association with The Open University

The Open University
Walton Hall, Milton Keynes
MK7 6AA

First edition published by Fracture Training Associates 2002.

This edition published 2009.

Edited and designed by The Open University.

Typeset in India by Alden Prepress Services, Chennai.

Printed in the United Kingdom by TJ International Ltd, Padstow

ISBN 978 0 7492 2370 0

2.1

The paper used in this publication contains pulp sourced from forests independently certified to the Forest Stewardship Council (FSC) principles and criteria. Chain of custody certification allows the pulp from these forests to be tracked to the end use (see www.fsc-uk.org).

FSC
www.fsc.org
MIX
Paper from responsible sources
FSC® C013056

Contents

Introduction

The blossoming of technology across Europe during the Industrial Revolution promised huge rewards for successful innovators and industrial spies – and there were no shortages of either! The promise of such rewards provided a driving force for the cross-fertilization of ideas, creating a cascade of benefits.

There is much fun to be had in arguing the relative merits of different innovations and their role in the development of industrial economies, but choosing one development in preference to others probably misses the synergistic point. In this book such pleasures are deferred in favour of the importance of metal technologies, starting with iron and steel, and in particular the understanding of how to predict the failure of an engineering part.

However, before a part can fail it has to be manufactured. Producing useful shapes in ferrous metals was, and still is, characterized by a fundamental conflict between those properties needed to make a shape and those which make that shape useful. Forming wet clay is easy, but the resulting pot is useless until it is fired and its properties changed in favour of performance. There is a similarly massive change in properties when molten metal is cast and frozen into a useful shape; irons and steels are most usefully formed and changed in their solid state to create pressure vessels, beams and drinks cans.

The interplay between the properties that favour manufacture and those that favour performance is subtle. A metal very easily rolled or drawn into tube will not resist everyday damage in service as readily as one that is difficult to form.

At the beginning of the nineteenth century this distinction was poorly understood. The mucky, hot chemistry of iron and steel making resulted in metals of varying quality and properties. In turn, evidence from failures was both common and confusing.

The most puzzling problem facing early engineers however was that their expensive structures, made from strong, hard steel, fractured routinely and unexpectedly. A successful tool for cutting metal should not blunt or chip readily, successful cannon should not burst and a mine chain should not break in service; but chip, burst and break they did, and in numbers.

From the start of the nineteenth century it took the best part of eighty years before a decent engineering toolkit of models and data was available to understand the failure and fracture of metal parts and structures. This text will explore the development of those ideas on engineering failure in the context of engineering imperatives from different industries at different times.

From mine machinery through railways, ships and aircraft to power stations, when new industries flourish they provide the driving forces that develop engineering methods and models. The unusual risks associated with warfare continually justify society's massive investments in technology over all of these industrial sectors.

For the first half of the nineteenth century it was assumed that a properly made metal part was flawless, and it was not considered polite to talk about cracks in fabricated metal structures that were bought and paid for. It is now assumed that even the most safety-critical metal products contain cracks from their manufacture and passengers routinely fly in aircraft containing growing cracks. This book is about living with cracks.

Living with cracks

Iron was the dominant structural material in Europe in the nineteenth century. By 1850, Britain alone produced two and a half million tons. Coal, for coking ovens, to power steam engines and blow air for blast furnaces, to draw wagons and drive ships, created an industrial virtuous circle. There was an insatiable appetite for structures and components made from iron and steel in all sorts of sizes, shapes and forms. This ferocious economy demanded new industrial knowledge to achieve its ends.

The railways and the mines used cast or pig iron, and its tougher product, wrought iron, in quantity. Accidents caused by unexpected fractures were common. Replacements were needed frequently, so testing became vital. Figure 1 shows a water wheel-powered fatigue test developed by the German mining engineer W. A. S. Albert in 1829. Figure 2 shows August Wöhler's machine for testing railway axles (Wöhler had published the seminal work on fatigue in railway axles in 1871). His work was triggered by unexpected fractures in an industry that was doubling in size every decade in the late nineteenth century.

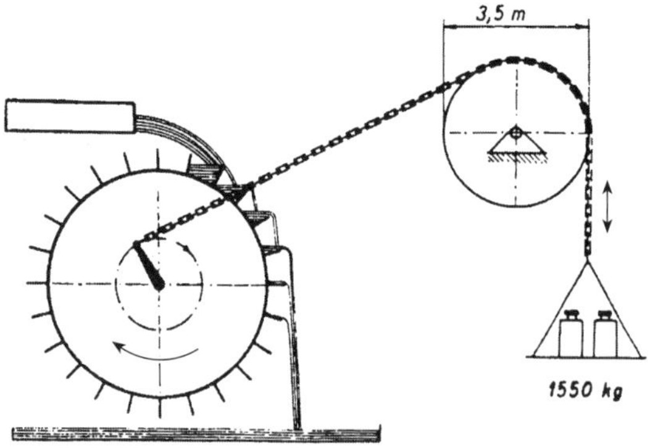

Figure 1 Fatigue testing using Albert's water-powered machine

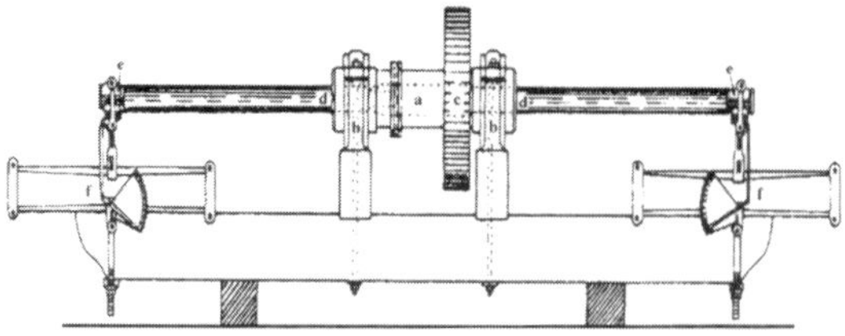

Figure 2 Wöhler's railway axle testing machine

European advances in iron and steel had gathered momentum in the eighteenth century. Folded and layered steel, famously used to make Samurai swords, was developed for tool and instrument making. This process was eventually surpassed by crucible steel manufactured by the clockmaker Benjamin Huntsman in the middle of the eighteenth century. Huntsman's method of steel production remained an important British monopoly for over fifty years. During this period it was the instrument-making industry that drove industrial developments, in particular fine tools and files, used to make accurate gears, wheels, slides and springs for navigational instruments, telescopes, microscopes and clocks. Until John Harrison's completion of his prize-winning marine chronometer in 1759, and its successful sea trials in the 1760s, navigation to the riches of the East was dominated by the hit-and-miss affair of following lines of latitude. The commercial and military rewards that followed from accurate navigation combined with naval force were enormous.

Figure 3 Bessemer in his 80th year

It had been known that carbon was critical to the production of high-quality steel since the 1770s. Hard steels for cutting edges and wear resistance were then the dominant industrial needs. These were satisfied by crucible steel produced in relatively small quantities. The production of steel was revolutionized by Henry Bessemer, Figure 3. Driven by the need to produce sufficient quality steel for large cannons, Bessemer developed William Kelly's idea of burning carbon out of carbon-rich cast iron.

> *At that time nearly all our guns were simply unwrought masses of cast iron, and it was consequently to the improvement of cast iron that I first directed my attention.*
>
> *(quoted in Hutton, 1996)*

By 1860, the Bessemer process was able to produce significantly larger quantities of refined steel than the crucible process. This development was so important that the *London Times* printed Bessemer's technical paper in its entirety. Commercialization of his and others' techniques paved the way for steel production of thirty million tons per annum by the end of the eighteenth century.

On the whole, large structures such as bridges, boilers and ships were made from components joined by riveting and bolting. If the behaviour of individual components could be understood then, by and large, understanding and analysing an assembly of individual components could solve any structural problem.

Chains, beams and plates that could be tested in the forms and sizes used in production were the stock-in-trade of heavy industry. However, the design and production of testing machines could never keep pace with the development and construction of huge structures. Testing machines reached their zenith with The Emery Machine, also called the US Testing Machine, shown in Figure 4. It was commissioned in 1879 at a cost of over $40 000 to the US taxpayer. In operation this hydraulic machine broke a 5-inch diameter bar at 361 tons, and a horsehair at a pound weight immediately afterwards. It remained in use until the 1960s. The Emery Winslow Scale Company still bears the name of its founder, A. H. Emery, and was still selling testing machines at the time of writing (2003).

Figure 4 Now that's what I call a testing machine!

Eventually, the ability to produce very large forgings and welded structures in high-strength materials outgrew the capabilities of all testing machines. The problem of relating the performance of small specimens to the behaviour of the same material in a massive structure then became vital to industry.

This link between sizes that could be tested and sizes in use was pushed to extremes by the war industries in a competition between weapons of attack and defence. Nowhere was this more evident than in the competition between armour plate and armour-piercing shells that eventually resulted in pocket battleships. *Dreadnought*, the first of the big-gun battleships, was built in 1905. It was equipped with a steam turbine to allow high-speed cruising without the problem of overheating main bearings that had plagued reciprocating steam engines, and had limited their time at speed.

The final act in the saga of the big-gun ships was played out nearly half a century later when the *Bismarck* sank the *Hood* using shells weighing over a ton, fired from ten miles away. The *Hood* was due to have its deck reinforced when next out of commission. However, before this could be done the plunging shells from the *Bismarck* penetrated its magazine.

Another illustration of the importance of being able to predict the behaviour of large structures was provided by the Liberty ships. They were cargo carriers, mass-produced in the USA during the Second World War. In benign conditions some of the Liberty ships broke in two and sank. The mass production of ships using welding fundamentally changed the behaviour of the ship structures. A ship's hull changed from being a collection of joined plates that could be individually tested to being a monolithic, single component that could not be tested, other than in use. To be strictly accurate the Liberty ships were not the first welded ships, but they were the first to be produced in large numbers: over two and a half thousand ships were launched between 1941 and 1945. The production record was five days for ships made by inexperienced welders. The quality of steel used and the manufacturing expertise employed varied considerably from yard to yard. This production rate compared with 245 days for a ship built by traditional methods. Henry J. Kaiser's mass-production systems were one of the most important individual contributions to the progress of the Second World War.

The quality of the ship steel in use at the time was very variable. It could fail in a brittle, fast fashion at, say, 0°C and in a ductile, slow manner at higher temperatures in the sizes used for hull plates. These ships experienced low temperatures and very high loads during the Atlantic winter convoys. However, some failures occurred under relatively benign conditions: the USS *Schenectady*, Figure 5, broke in two in calm waters after only one day of sea trials. Its deck stress at the time of fast fracture was the very low value of 9900 psi (68 MPa).

A welded ship's hull is a long, hollow beam supported on wave peaks. It can experience high, tensile stresses due to cargo loading, even when the load is well distributed along the keel. In many of the Liberty ships cracks grew rapidly from sharp corners at hatch openings. In extreme cases this caused the hull to break completely in two. Design changes were introduced during the spring of 1943 to reduce the sharpness of hatch corners, incorporate crack-stopping devices, and improve steel and welding procedures. These modifications caused the failure rate to fall dramatically.

Figure 5 USS *Schenectady* 'Type of failure: broke in two'

Unfortunately, similar failures do still occur. The SS *Bridgewater* broke in two in the 1960s in the Indian Ocean. The MV *Derbyshire*, a massive 'Capesizer' of over three football pitches in length, almost certainly broke in two during a typhoon off Japan in 1980. Evidence suggests that the failure was caused by fast fracture; she sank without sending a distress signal.

Aircraft, designed to be light enough to fly, are still riveted together from thin, aluminium sheet and extrusions that transfer load to more substantial forgings and machined parts. The consequence of a highly stressed structure made from a light material is that cracks grow during service under the cyclic loadings of take-off, flight and landing. Boeing's commercial aeroplanes are designed with an economic structural life objective of 20 years in service with 95% reliability.

> *The economic life objective is for 19 out of 20 airplanes to exceed 20 years of operation without major fatigue cracking in the primary structure. For structure that just meets the design objective, 1 out of 20 airplanes may crack prior to 20 years of service. In a large fleet of airplanes, the earliest fatigue cracking could occur as early as midway through the design life, even though most of the fleet will exceed the full design-life objective without cracking.*
>
> *(Hall and Goranson, 1982)*

Since the 1960s commercial planes have been designed and certified according to a fail-safe design philosophy. This means that wherever possible, alternative load paths are designed to take up loads shed from a failed component. For example, wings are designed to bear the full design-limit load with a crack in the skin extending across two stringer bays (the stiffening sections within the wing). Aircraft structures have become more 'distributed' over the years, in the sense that single large beams, called spars, are less obvious in a design where more members carry primary loads.

Where this design philosophy is not possible, for example in steel undercarriage legs, a safe-life approach remains in use. Safe-life means that a part is thrown away after a predetermined number of cycles in service. Military aircraft are designed to operate at higher peak stresses and with a wider flight envelope than civil aircraft. Your holiday charter, for example, is not designed to be flown upside down. The higher performance requirements mean that military aircraft cannot afford to carry extra weight around in the form of additional, or fail-safe, structure. They are therefore designed on a safe-life basis. The differences in flight stresses endured by military aircraft, even of the same type, require that component loadings are extensively monitored in service. Specific components on individual aircraft have their accrued fatigue damage recorded.

The F-111 fighter aircraft was designed in the 1960s. Its wings are designed to pivot about huge pins so that they swing forward for low-speed manoeuvring and are swept back for high-speed flight. In the late 1960s, after an aircraft lost a wing in the air, it was discovered that the critical crack size in the high-strength steel used was close to the smallest detectable flaw size. Low-temperature ground testing under maximum loadings caused several further wing failures. This is a classic example of buying high-strength, and in this case reasonably good fatigue properties, at the expense of low fracture toughness. This problem led to the use of damage-tolerant design using fracture mechanics on critical, fail-safe components.

In the 1980s the fail-safe approach developed into damage-tolerant design for commercial aircraft. This philosophy introduced structural inspection procedures into the (safety) type certification of an aircraft. It demands a high probability of detecting damage in a part before its strength deteriorates to below the regulatory limits. In practice this means inspection engineers, armed with large compendiums of crack shapes, decide whether a growing crack needs to be dealt with immediately or can wait until the next service.

Adjacent, interacting flaws caused the notorious accident on 28 April 1988, shown in Figure 6 and on the front cover. An Aloha Airlines Boeing 737, shown on landing in Hawaii, suffered massive structural failure at 24 000 feet. This accident stimulated the developing interest amongst both the civil and military aircraft communities in applying damage-tolerant ideas to interacting flaws. However, the probability of the large pressurized fuselage of a civil aircraft containing interacting flaws, all under high stress, is much greater than that in an unpressurized military aircraft.

Figure 6 A Boeing 737 unzipped

By way of contrast, the thin, sheet steel used in motorcar production is very lightly loaded in normal use. The critical design parameters in this industry concern a structure's formability by press tools and energy absorption during a high-speed crash, rather than the understanding of crack growth in service.

The high-toughness steels used in automotive production are about five times the toughness of the aluminium alloys used in aircraft. In thin sections car body steels deform plastically under extreme loads without tearing. The design stresses are so low that fatigue crack growth in the body is not a routine problem.

Suspension and rotating parts for cars are designed for a notionally infinite life; the durability of the modern car is a tribute to the success of this approach. An observable fatigue crack in a test is deemed to be a failure. In addition, there is no need for crack growth monitoring or part replacement (except for timing belts, which are replaced routinely and are therefore considered safe-life designs). An ill-fated attempt by the manufacturer GKN to introduce glass-reinforced plastic (GRP) springs for goods vehicles would have required these 'cart' springs to be replaced as part of routine servicing. The unsprung weight saving was a significant performance advantage, but the difficulties of mass production quickly killed the design and closed the factory.

These arguments suggest that there is significant potential for performance improvements that lead to weight saving in vehicle design. However, for the moment, the important structural parameter for a car is stiffness and, whilst weight saving is important, parts with adequate stiffness are usually more than strong enough. The guiding principle in car design is crash protection of the occupants by crumpling of the structure. However, consider the example of nuclear pressure vessels, where safety is paramount. They are constructed from massive open die forgings, Figure 7, welded together for the reactor pressure vessel, and machined for the generator rotors.

Figure 7 The scale of a Sheffield Forgemasters product

The reactor vessel of a gigawatt pressurized water reactor (PWR) for example, is made from six forgings that weigh a total of 500 tonnes. The ingots that are the basis of the forging are two to three times the weight of the finished part. The heaviest forging is the low-pressure rotor, which weighs over 200 tonnes. It's hard to imagine the effort involved in the manipulation and deformation of such pieces.

The very high toughness, manganese–nickel–molybdenum steels used to make a nuclear vessel contain cracks and defects introduced during welding. The massive size of the forgings means that these cracks are heavily constrained by the surrounding metal. The rejectable defect size in the crotch of a nozzle for the Sizewell 'B' nuclear reactor was 5 mm at manufacture, compared with a calculated, critical defect size of about six times that length under the most onerous imaginable loading condition.

There is no fallback position if a nuclear vessel fails. Therefore before a nuclear plant can be operated, a safety case must be made that demonstrates that failure is an 'incredible' event with a frequency of occurrence of less than 10^{-7} per annum. This is two orders of magnitude less than the demonstrated failure frequency of conventional pressure vessels. These figures are equally hard to imagine. So, as with the aircraft industry, the nuclear industry has to live with cracks.

Accepting the inevitable presence of cracks and the likelihood of their growth during service leads to industrial methods for the calculation of service lifetimes. One often hears of industrial plant or aircraft having an expected service life of 'so-many years', but it is a mistake to think of this as immutable. The initial assumptions of a design or failure assessment can be revisited and lifetime calculations can be revised. For example, there is a crack in a safety-critical part of the Dinorwig pumped storage system that is 3 mm long. Every time that the part is inspected and the length of the crack measured, calculations about the expected lifetime of the part are revisited. If the crack has not grown, the 'clock' is reset, i.e. the potential life of the item is extended. This particular crack is so well known that it is surprising it has not been named.

Similarly, aircraft originally designed within one paradigm, for example safe-life, can be revisited using newer procedures, such as fail-safe or damage-tolerance, and their operating profiles and lifetimes recalculated.

The highly regulated nuclear industry had to develop its engineering theory and practice at the same time as building and running the plant. When US pressurized-water reactor designs were first used in the UK, the power plant operator, the Central Electricity Generating Board (CEGB) took over the development of the reactors' design, inspection and validation criteria. The economic and technological effort needed to achieve this was immense. During the 1970s and 1980s formal procedures for assessing the integrity of structures containing defects were developed by the CEGB because there was no suitable guidance from other industries or institutions. These procedures are such powerful representations of failure space that they continue to evolve at the time of writing.

To fully understand current industrial practice requires the history of engineering failure and fracture to be outlined. Before a part can fail it has to be made, so this is where the story starts.

Ductility and the making of shapes

It was an obvious first step to try to shape naturally occurring gold. Those metals that could be shaped by hammering, bending and twisting were highly prized for jewellery making and decoration a long time before metals were refined from ore and hardened for weapons. Gold, found as nuggets, could be used without refining and was discovered to be extraordinarily malleable.

> *Gold ... by reason of a faithfull tenacity and ductilenesse, will be brought to cover 10 000 times as much of any other Mettall.*
>
> *(Donne, 1644)*

Gold leaf, of perhaps a tenth of a micrometre thick, can cover about 20 m^2 from 30 g of gold. This valuable material can be used for decorated letters on high-quality book bindings.

A good definition of malleable that uses the term in association with fracture, is:

Malleable

Having the property (possessed by certain substances, esp. metals) of being deprived of form by hammering or pressure, without a tendency or capacity to return to it, or to fracture.

After foil and sheet came wire for decoration:

> *And they did beat the gold into thin plates, and cut it into wires, to work it ... in the fine linen.*
>
> *(Exodus, 39)*

Wire, whether for decoration or for more mundane use, was made by this 'slit and hammer' method until 400 AD, when wire drawing through stone dies by lunging or using winches, swings and ratchets was developed. Wire drawing consumes a lot of energy, Figure 8.

The term 'ductile' has long been commonly used to describe metal plasticity in the context of wire drawing.

> *The other sort of copper ... yeeldeth to the hammer and will be drawne out, whereupon some there be who call it Ductile*
>
> *(Holland, 1601)*

> *All Bodies Ductile (as Metals that will be drawne into Wire)*
>
> *(Bacon, 1626)*

Figure 8 Wire drawing through a die

The plasticity of metals is the basis of forging useful shapes, whether on an anvil or between closed dies. Deformation is engineered to take place under compressive stresses, so preventing tensile stresses from creating and opening cracks.

> *Science nor Crafft to hym was delectable, but to forge malyable mataylle.*
>
> *(Lydg & Burgh, 1450)*

Although the terms malleable and ductile were used interchangeably:

> *Metals are malleable or ductile under the hammer.*
>
> *(Sullivan, 1794)*

later definitions distinguish between malleable and ductile:

> *All the metals, that have been described as malleable (with the exception, perhaps, of nickel) are also ductile, or may be formed into wire.*
>
> *(Henry, 1826)*

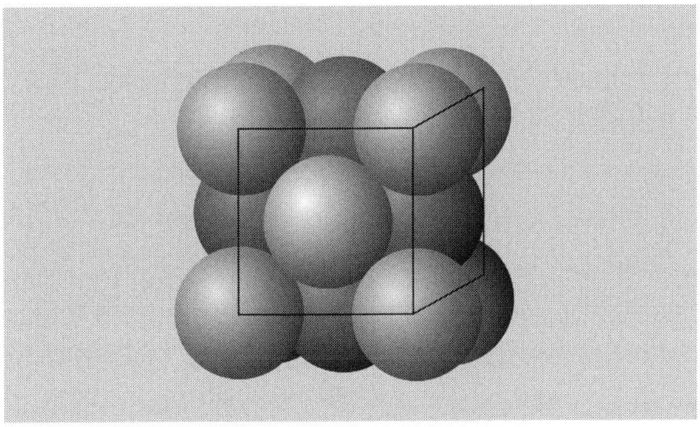

Figure 9 Packing atoms closely in a face-centred cubic array

The fundamental physics of metal plasticity governs both ductility and malleability. Gold and aluminium, for example, are face-centred cubic metals, Figure 9. They have more slip systems (slip planes and slip directions) compared to body-centred cubic or hexagonal packing, so they work easily under the action of shear on a close-packed plane.

In 1611 Johannes Kepler, now best known for his heliocentric (Sun-centred) model of the solar system, experimented with the problem of packing spheres. He concluded that the arrangement known as face-centred cubic packing, a pattern favoured by fruit sellers, could not be bettered. It was not until 1998 that Thomas Hales of the University of Michigan announced a 250-page, 3-gigabyte computer-based proof of what has become known as The Kepler Conjecture, explaining why this is the case.

Slip in metals can be pictured as waves of atomic movement spreading out over a plane. The boundary between a slipped and an unslipped region, as one grows at the expense of the other, is called a dislocation line: it is a configuration of displaced atoms. Dislocations are the 'carriers' of plasticity in metals.

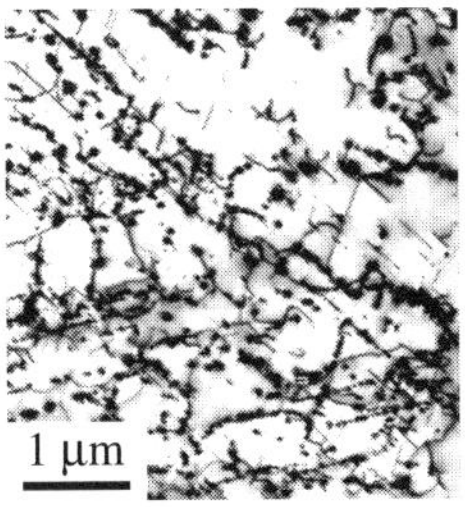

Figure 10 Picturing dislocations in foil

Dislocations were first suggested in the 1930s and became widely discussed during the 1940s. The metallurgical community showed great resistance to accepting this phenomenon until electron microscopes provided crucial experimental evidence of the presence of these line defects. They were seen moving under the thermal stresses caused by electron-beam heating of thin aluminium foil (Figure 10) when its hardening was being studied at the Cavendish Laboratory (Cambridge, England) in the 1950s.

Alloying and heat treating of metals gives the engineer fine control over ductility, and therefore strength and toughness, by the microstructural mechanisms of solid-solution, precipitation and dispersion strengthening. These very different mechanisms all control dislocation mobility in crystal lattices. No such unifying ideas were available to inform metallurgists designing new alloys and heat treatments at the start of the nineteenth century, nor were the means available for observing plasticity in metals at a microstructural scale.

However, plasticity can be characterized at a macroscopic level. The engineer's standard test for plasticity in metals is the uniaxial tensile test. A bar is pulled apart until it fails and a plot of load against deflection records the event.

In order to generalize the results of a tensile test, the load is divided by the bar's cross-sectional area, and the extension is divided by the bar's length, so producing derived properties named stress and strain, respectively.

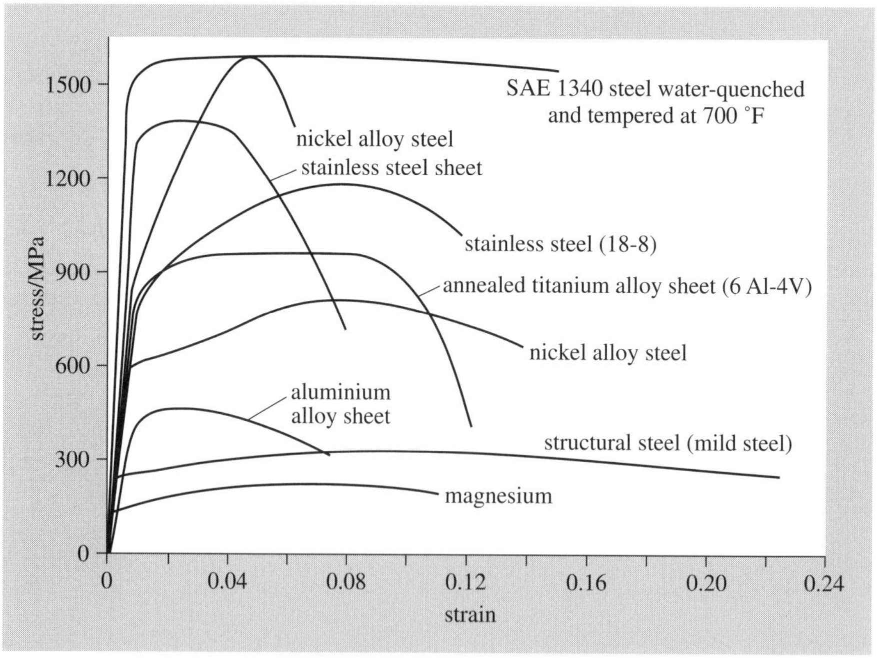

Figure 11 A variety of stress–strain curves

Plotting the stress–strain curves of modern metal alloys produces a wealth of different shapes, Figure 11. All of these metals show an initial linear response during which deformation is recoverable – all of the internal displacements return to zero – if the load is removed. However, at some value of stress a metal yields and subsequent deformation is not recoverable. To avoid manufactured metal parts changing shape permanently during use it is sensible to define the stress at which this happens as the yield stress of a material, and ensure that parts are designed to be loaded well below yield.

It is not generally easy to identify a particular point where yield occurs, σ_{yield}. The '0.2% proof stress', a pragmatic value of stress that gives a permanent (plastic) strain of 0.2%, is specified as yield for general engineering purposes: $\sigma_{0.2}$. Metals are considered to be highly stressed if they operate at about two-thirds of the yield stress. Above the yield stress, useful metal alloys strengthen by work-hardening mechanisms that inhibit dislocation movement until the ultimate tensile stress, σ_{UTS}, which is the maximum value of stress in a tensile test, is reached. Beyond σ_{UTS}, a local neck forms in the bar and so the stress, measured as load/area in the tensile test, reduces until the test piece fractures. This is because stress is measured by engineers using the nominal cross-sectional area of the bar and not the reduced area at the neck. All metals eventually fracture in a tensile test; a die is needed to suppress necking when wire is drawn.

The variety of stress–strain curve shapes pays tribute to the metallurgist's success in creating metal alloys. Some alloys work harden strongly to high values of stress and some work harden gently to high values of strain. In general there is a trade-off between strength and ductility in the choice of a metal alloy to create a work-hardening response appropriate for use.

The approach to the modelling of work-hardening depends on the engineering context. A metal forger, who wants extensive plasticity, will have little interest in, and might well ignore, elastic deformation and take an average of σ_{UTS} and σ_{yield} (called the flow stress, σ_{flow}) and presuming it to remain at that stress over the entire strain range: this is a 'perfectly plastic' model. Introducing the linear elastic portion of the curve together with the flow stress approximation, or possibly using yield stress, creates an elastic–perfectly plastic model.

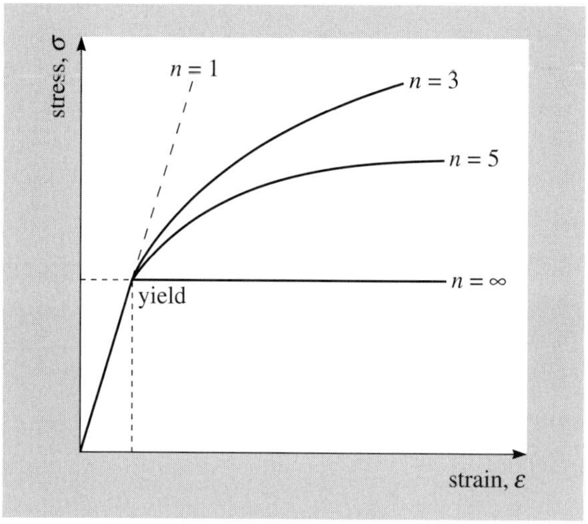

Figure 12 Modelling work hardening

If the work hardening of a metal needs to be modelled mathematically a power-law relationship can be used to approximate the shape of a curve up to necking, Figure 12.

$$\varepsilon = \frac{\sigma}{E} + \alpha\sigma^{n}$$

where

α = work-hardening coefficient, and

n = work-hardening exponent.

This is the Ramberg–Osgood relationship. For a strongly hardening material a value of n might range between 3 and 5, whereas a gently hardening material might have a value of n as large as 20.

Plasticity plays a key role in determining the load at which a failure occurs and the type of failure that takes place in an overloaded part. A dictionary definition of ductile links the words 'brittle' and 'tough':

Ductile

1. Of metal:

a. That may be hammered out thin; malleable; flexible, pliable, not brittle. Still frequent in literary use: for technical use, see b.

b. Capable of being drawn out into wire or thread, tough. [The current technical use.]

Here I shall use the words brittle and tough (or ductile) to describe a fracture rather than to describe a material, which is the more common usage. I shall demonstrate that many engineering metals can fail in either a brittle or a tough way, depending on the size and shape of the engineering part into which they are made.

Early irons and steels were extraordinarily 'mucky' by modern standards. For example, a major problem with the industrialization of the Bessemer process was the need to remove phosphorus from the ore. Phosphorus segregates at grain boundaries, embrittling the steel by providing an easy route for a growing crack.

One of the major influences on the performance of modern steels is still cleanliness. Stirring iron ore, coke and chemicals in a pot exposed to atmospheric gases at over 1500 °C is not an inherently clean business!

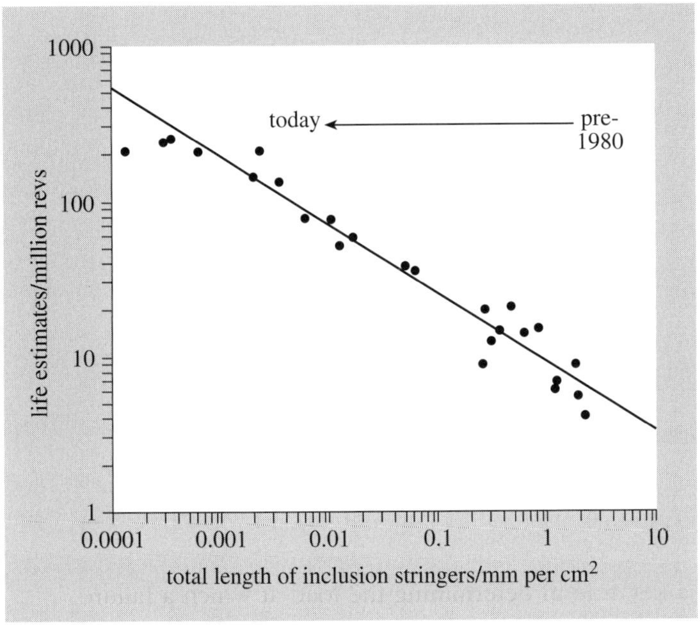

Figure 13 Improving the performance of bearing steel through cleanliness

The graph in Figure 13, published by Timken [the bearing makers], shows how the performance of bearings has improved markedly over time by reducing the amount of inclusions left in a steel after processing. Note the units of cleanliness shown on the diagram: length of inclusion stringers per volume of steel. Very clean steel is also required for the manufacture of steel drinks cans to draw down wall thicknesses to those needed to challenge aluminium in the marketplace.

As well as the uncertain chemical make-up of early steels, the quality control of the production and treatment of components was also hit and miss. The large amounts of experimental data generated from engineering failures made interpretation difficult.

However, what was recognized was the important effect of changes in section, and of notches, on how an iron or a steel part fails under load.

Notches and the breaking of shapes

The industrial context during the first half of the twentieth century became bewilderingly complex. Parts were manufactured from rapidly developing materials of different qualities and compositions, in shapes containing screw threads, holes and corners, under complex loadings at different temperatures. From a modern perspective it is difficult to imagine the confusion of that time, and the consequent proliferation of explanations for failure and fracture.

Two observations are key to this narrative. First, that a bar notched to different degrees of sharpness is a good model of practical shapes and forms. Second, at that time defects such as welding cracks were assumed to be absent in properly produced structures. It is now accepted that even safety-critical structures, such as nuclear pressure vessels produced under the most stringent manufacturing procedures, routinely contain welding cracks and forging flaws. It was not until the 1960s that cracks became of interest to engineers. Even then it was deemed impolite to discuss pre-existing cracks in properly fabricated structures. However, no one involved in designing loaded shapes could ignore notches. Depending on shape, material, size and temperature, different types of failure associated with notches were observed to occur in practice.

There were two extreme types of failure at a cross-section that had been reduced by the presence of a notch: fast, dangerous fracture caused by a crack growing from the notch, or slow plastic collapse of the section with no crack growth. Against this background of radically different physical events the imperative to test notched metal was irresistible.

Small bars of steel containing slots and notches of various types are still routinely tested by breaking them with a pendulum hammer, Figure 14 overleaf. The test piece breaks in two, so removing energy, which is assumed to be proportional to the work of fracture, from the swinging hammer. This energy absorbed during fracture is measured by the angle to which the hammer swings after breaking the bar at the notch.

Pendulum machines, Figure 15 on page 25, are cheap, simple and quick to use for making measurements at a range of temperatures. Such tests are carried out according to the procedures of international standards that capture custom and practice to ensure that results are comparable across test houses in different countries. Swinging hammer tests have been carried out for over a hundred years on specimens of varying sizes and severities of notches.

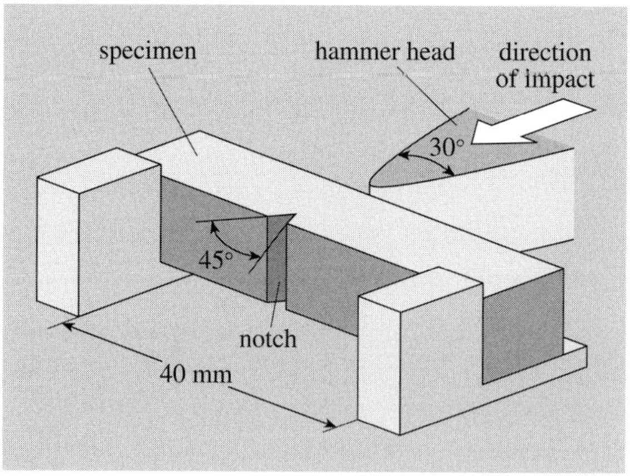

Figure 14 The detail of a pendulum hammer test

The importance of ductility had been recognized by Considère in 1889. He attempted to create a measure of 'fragility' as the difference between the ultimate tensile strength and the proof stress of a metal, or as the ratio of these two properties. The physical driving force for formalizing such a measure was the association between a lack of ductility and the ease of fracture of a metal:

> *One has always known that certain metals are regularly fragile, that is to say, that they always break, without practically undergoing any extension, and that consequently their rupture requires only a very small expenditure of work.*
>
> *(Le Chatelier, 1904)*

At the beginning of the twentieth century the use and interpretation of different materials properties and types of notched bar impact test were of great interest. They were also the source of fierce arguments.

> *Impact tests and especially impact-bending tests on notched bars have generally been regarded as a means of determining the fragility of metals. But people have not sufficiently preoccupied themselves with exactly defining the meaning of this term, which certainly has been abused in so far, as some authors simply state that certainly metals are fragile and others not fragile, whilst most of the devices applied aim at arriving at some universal value.*
>
> *(Charpy, 1909)*

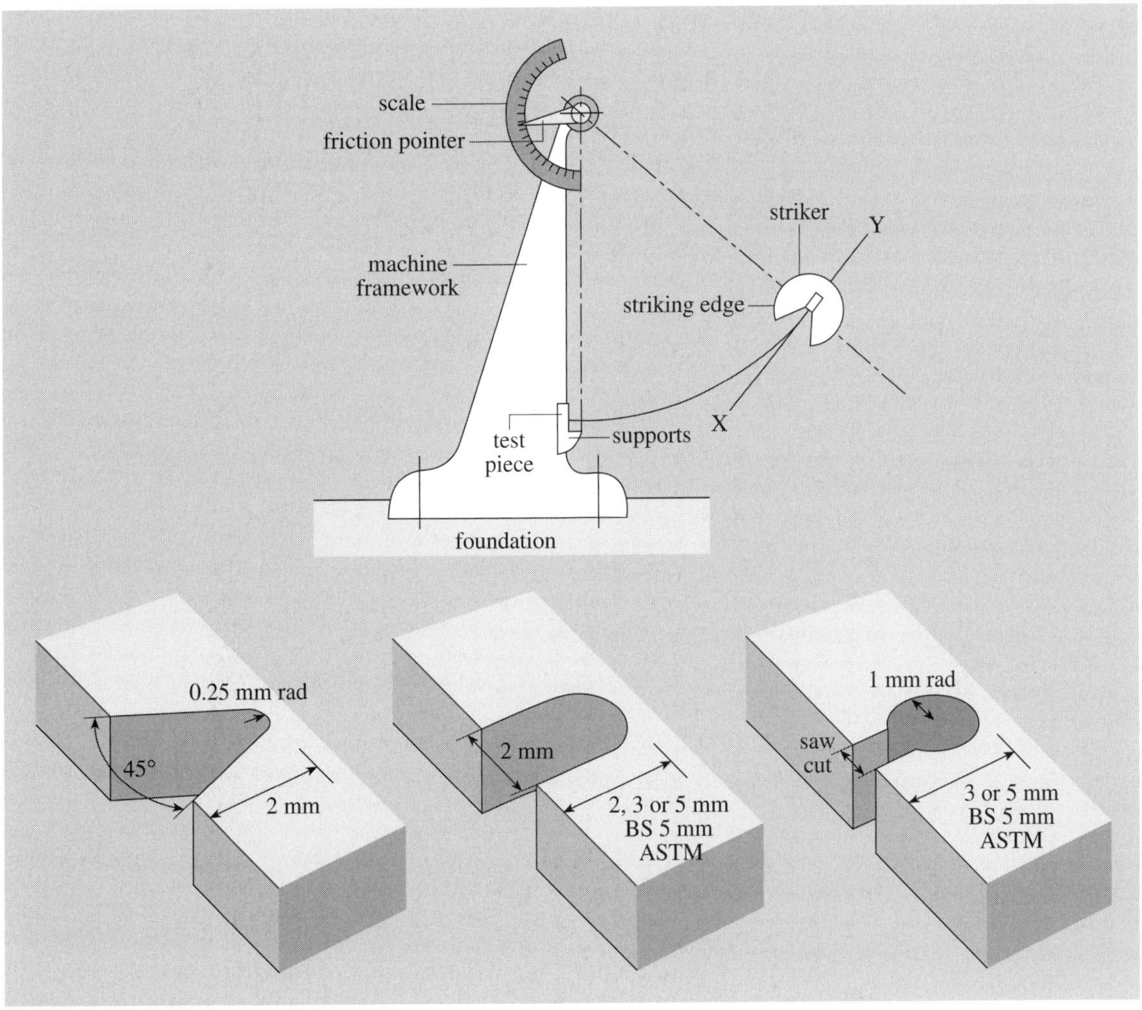

Figure 15 The swinging hammer test and some specimen notch geometries

These early experimenters knew that there was a relationship between the deformation before fracture and the work absorbed during fracture. Therefore their problem was which to choose in order to characterize the 'fragility' of a metal.

High loads were most easily produced by falling weights. This sparked a debate about whether measurements from impact loads produced different results from static loads.

Figure 16 Augustin Charpy

Figure 16 shows Augustin Charpy. In addition to a brilliant industrial career, culminating in his appointment as Director General of the *Compagnie de la Marine et Homécourt* in 1920, Charpy was one of the founders of the science of alloys in France. He was elected to the *Académie des Sciences*, in the newly created section of the industrial applications of science, and was a highly regarded professor at the *École Supérieure des Mines* in Paris.

He performed experiments with weights dropped from 47 m high chimneys ('fast loading'). Results from these tests, in conjunction with slow loadings convinced him that a falling weight was simply a convenient way of applying a high load 'within the limits which we can attain with the fall of a tup'. It is now known that typical loading rates in a modern Charpy machine are about a tenth of that needed to cause reductions in the toughness of metals.

Intuitive ideas that are wrong are extremely difficult to displace. As well as wrestling with the idea that a hammer blow looked obviously different, and in some way more severe than a static load, engineers at the beginning of the twentieth century were still debating the notion that metals failed because they had 'crystallized'. The sparkling appearance of a brittle fracture was still seductive, even though the idea had been refuted by William Rankine, working on railway axles in 1843, and by David Kirkaldy, working out of the Scottish shipyards in 1861, who observed:

> *that a fracture surface could have either a fibrous* [the term 'dimpled' is now used to describe a ductile failure surface] *or crystalline appearance, solely by altering the shape of the specimen.*
>
> *(Kirkaldy, 1862)*

This very important geometric effect governs much of current thinking on the fracture of metals. Utility, argued Charpy, was a key notion to the development of a practical fracture test that did not require tall chimneys.

> *The apparatus disposed in such a manner as to measure the energy remaining in the hammer after fracture, gives this value [work done in fracture] directly, while the use of a testing machine for slow tension depends on stress–strain diagrams, whose areas have to be evaluated by the aid of the planimeter which is a somewhat tedious and delicate operation.*
>
> *(Charpy, ibid.)*

E. G. Izod, promoting the impact test named after him, wrote:

> *It can readily be seen that in the use of these very small test pieces lies one of the chief merits of the whole system, it being possible to cut specimens from crop ends of shafts, odd scraps from machinery and very often from the finished product without injuring its strength or appearance, further the cost of preparation of the specimen is very small, and storing and recording of the fractured sample for future reference is greatly simplified.*
>
> *(Izod, 1903)*

Charpy recognized the importance of the notch as a stress concentrator and described it as:

> *The element which forms the bottom of the notch will much more rapidly be strained than all the others and will already be broken, when others are scarcely strained yet.*
>
> *(Charpy, ibid.)*

He found that notched-bar impact tests carried out under bending were better at differentiating between metals than notched-bar impact tests carried out under axial loading. He could not explain this, but chose bending because it could 'furnish information which other tests do not give'. Interestingly, in Charpy's published results he records eight materials' properties, but does not give the cross-section of the specimens and the metals tested are identified only as A, B, C and D, which is far from sufficient to understand the results from an informed perspective.

Before 1870 the structural iron alloys in common use were cast iron (including malleable iron), wrought iron, and crucible steel. After that date Bessemer steel, which is embrittled by cold work, and from which rails were made, became common. After 1890 strong, tough alloy steels became available for heavy construction; it was this development that drove the search for a means with which to measure toughness.

Not surprisingly, Charpy was keenly interested in relating the behaviour of metals under test to the behaviour of metals in use. His was a world of massive forgings and armour plate for ships; the French navy routinely used notched-bar bending tests for quality control.

The commercial benefit of this work was startling: as a result of quality-control tests heat-treated steel, used to make 8-ton steam hammers, lasted over six years rather than failing six times a year. In 1909, Charpy concluded:

> *The mechanical tests of metals present us with a great number of scientific and technical problems which are far from being solved. To utilize the tests in industrial practice it is sufficient to regard these tests as means for the identification and classification of metals.*
>
> *(Charpy, ibid.)*

The Charpy test is still in widespread use for these purposes. Attempts to correlate Charpy V notch energies with modern fracture toughness measurements bear testament to his judgement.

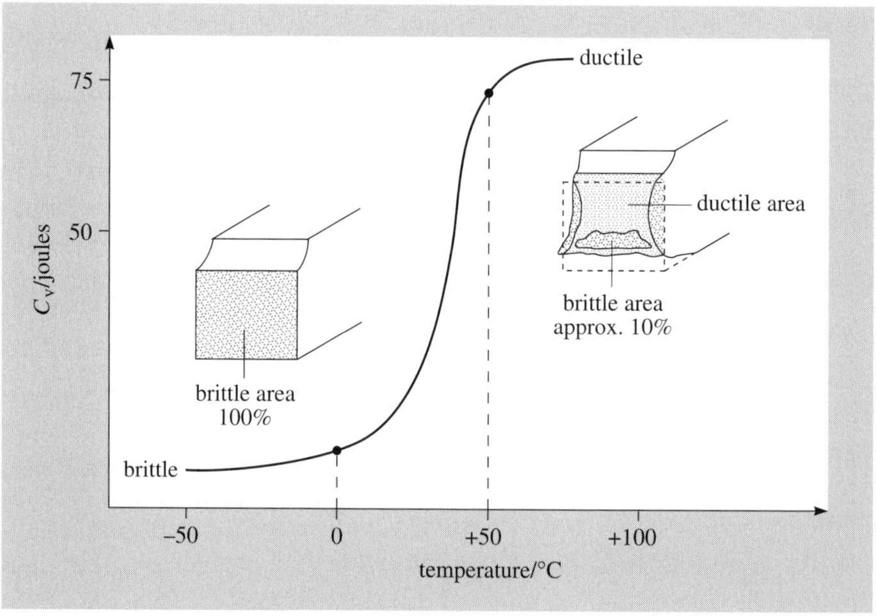

Figure 17 Mapping fracture from Charpy data

Results from Charpy tests are most comprehensively shown as a plot of energy absorbed during fracture against the temperature of a test, Figure 17. For ferritic steels, this classically shaped curve shows the transition from a brittle fracture on the 'lower shelf' to a ductile fracture on the 'upper shelf', designated as Charpy V notch energy or C_V.

This failure map shows how the nature of the fracture surfaces changes as C_V falls dramatically with reducing temperature. Brittle fracture surfaces are rough and planar, whereas ductile fracture surfaces are dull and dimpled, showing large deformation (shear lips) at the edges of the fracture. In between, surfaces show a typically ductile appearance surrounding some brittle failure.

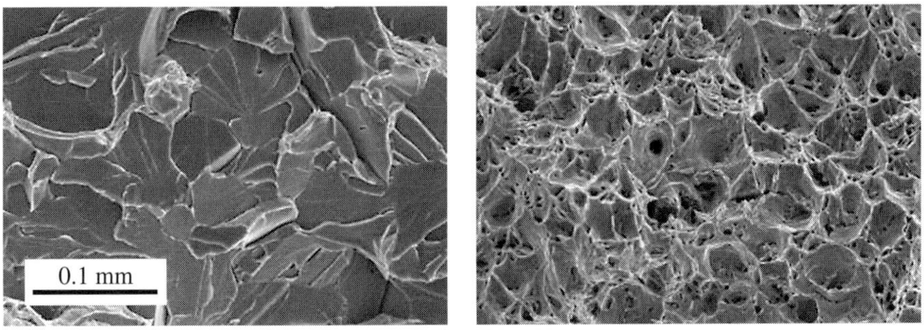

Figure 18 The detail of a brittle and a ductile fracture surface

Close inspection of fracture surfaces indicates that brittle, fast fracture cleaves through the ferrite grains of a steel, and ductile rupture tears the surface creating dull dimples over the specimen's surface, Figure 18. On this evidence, it is easy to accept that the ductile mechanism absorbs much more energy than the brittle mechanism.

Steels can be produced in either ferritic (body-centred cubic) or austenitic (face-centred cubic) atomic forms. There is a striking difference between the fracture behaviour of austenitic and ferritic steels: austenitic materials never cleave, whereas ferritic materials can show a surprising mixture of cleavage and dimpled rupture.

Egon Orowan, working at Massachusetts Institute of Technology, describes this well:

> *The fracture behaviour of ferritic steels is not merely an intermediate case between full brittleness and unlimited ductility. It represents a strange duplicity, an instability that can change the highly ductile behaviour seen in the common tensile test to almost complete brittleness.*
>
> *(Orowan, 1949)*

The sharp edges of a perfectly cleaved surface encourage the assumption that plasticity plays no part in the cleavage process. However, in 1945 Orowan performed elegant experiments, using X-ray scattering to investigate plasticity below the cleavage plane. This work on fractured ship steel showed that plastic strain is the dominant cause of energy loss in steels, even for low-energy cleavage fractures. Orowan did no further work on fracture, as this biographical quote explains.

Orowan did not pursue the study of ductile fracture and its mechanisms, and had very little liking for the formal developments in fracture mechanics initiated by Irwin, that were revolutionizing the study of fracture in engineering.

(Nabarro and Argon, 1996)

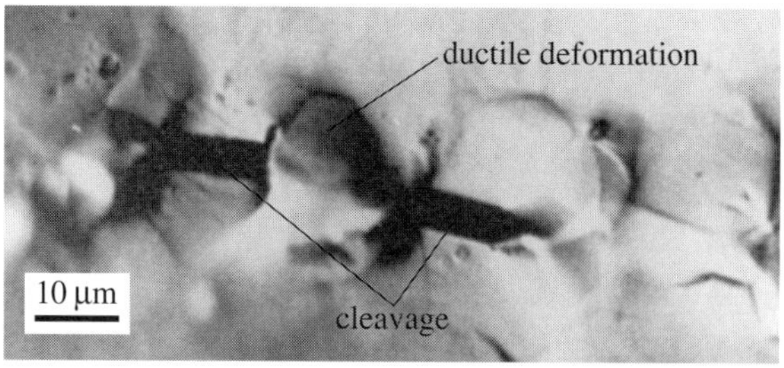

Figure 19 A deformed grain between two cleaved grains

It is now known that the puzzling fracture behaviour of ferritic metals is caused by the combination of a strongly temperature- and velocity-dependent yield stress, with a cleavage strength that is only slightly higher than the yield strength. So, cleavage and ductile rupture can coexist on the same fracture surface, Figure 19.

To an engineer dealing with a load-bearing structure that operates over a range of temperatures, a Charpy map provides a powerful tool for choosing competing materials, for controlling the quality of supplied materials and for understanding how a structure might fail. However, to move from qualitative interpretations of fracture and quality control of metal supply to predictive capabilities required more than a measure of the energies absorbed by different sharpness of notches and descriptions of fracture surfaces at high magnifications. It required a model of stresses in complex geometries, and an appreciation of how they might drive a crack through metal.

Stress concentration

In 1913 C. E. Inglis demonstrated how stresses were concentrated around sharp localized changes in shape. He modelled holes and notches in uniform stress fields to produce a two-dimensional stress concentration factor, k_t, of the form:

$$k_t = \frac{\sigma_{\text{peak}}}{\sigma_{\text{nominal}}} = \left(1 + 2\sqrt{\frac{a}{\rho}}\right)$$

The length of an edge notch is a and that of an embedded ellipse is $2a$. The radius at the tip of the notch (or root radius), modelled as an ellipse, is ρ. The simplicity of Inglis's equation, together with the quantity of corners and notches manufactured, guaranteed its popularity. Although the resulting equations are simple in form, they are not at all simple to derive: one guesses a complex stress function and works in elliptical coordinates. Experimental evidence and comfort for those of us not skilled with stress functions is provided by photoelastic tests.

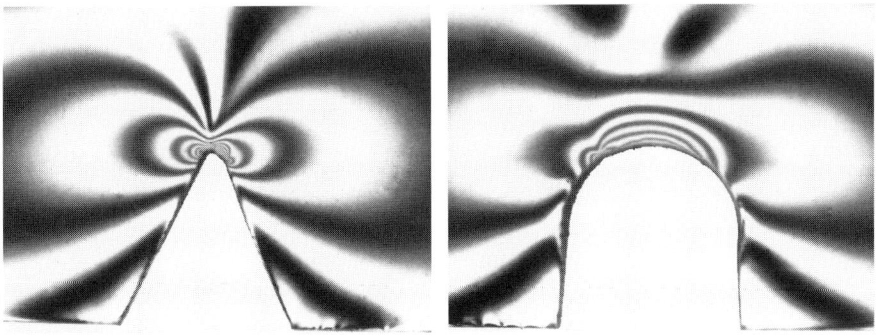

Figure 20 Stresses are more concentrated at sharp notches

The two photographs shown in Figure 20 illustrate how differently shaped notches concentrate stresses at their tips, and the shape of the stress distributions. The black fringes map regions of constant shear stress. The more the stress is concentrated, the greater the number of fringes that are shown by light shining through the photoelastic plastic. The sharp notch can be seen to concentrate the stress much more than the blunt notch.

The stresses in a photoelastic plastic are elastic, which means that the deflections return to zero when the load is removed, so they are unsuitable for modelling metal plasticity. However, they do provide an idea of the shape of stresses at a sharp notch. In addition, they demonstrate that necessary notches, such as screw threads, will generate much higher stresses than the nominal stress on an engineering part.

In the following decades Inglis's approach was extensively worked to produce stress concentration factors for all sorts of geometries, under all types of loadings. It is easy to put some back-of-an-envelope numbers into Inglis's equation in order to show how he provided engineers of his time with an important tool.

The peak stress at the tip of a notch in an engineering part designed to work at the moderate stress of a half the material's yield stress is calculated as:

$$\frac{\sigma_{\text{peak}}}{\sigma_{\text{nominal}}} = \left(1 + 2\sqrt{\frac{a}{\rho}}\right)$$

$$\sigma_{\text{peak}} = \frac{\sigma_{\text{Y}}}{2}\left(1 + 2\sqrt{\frac{a}{\rho}}\right)$$

where σ_{nominal} = the stress away from a stress concentration

σ_{Y} = yield stress

Choosing an innocuous geometry, perhaps a screw thread, modelled as a 5 mm long notch with a root radius of 1 mm, produces a peak stress of:

$$\sigma_{\text{peak}} = \frac{\sigma_{\text{Y}}}{2}\left(1 + 2\sqrt{\frac{5}{1}}\right) = 2.7\sigma_{\text{Y}}$$

So, in the presence of a notch, stresses above yield commonly occur at low loads in everyday geometries.

Now model the 5 mm long notch with a tiny root radius of 10^{-5} mm and Inglis's model produces an elastic stress concentration of well over a thousand.

For a sharp crack modelled as an ellipse with an infinitely small tip radius, Inglis's elastic analysis produces an infinitely large stress. It doesn't much matter to the Inglis equation whether the crack is 10 mm long or 15 mm long, which flies in the face of common experience.

However, remember that this is an elastic analysis. It is used to predict a numerical value of stress. It does not model the non-linear, irrecoverable plastic response of a metal stressed to above its yield point. In metals, what happens physically is that the material at the tip of the notch yields, so causing the stresses to redistribute and the sharp notch to blunt as the load is increased. This yielded material work hardens, so causing material further out from the notch plastic zone to yield, and so on. The detail and extent of this plasticity depends on the shape of a metal's work-hardening curve: a rapidly and steeply hardening material behaves differently from a slowly work-hardening material.

These notions explain the phenomenon of 'shakedown'. This is why metals can be used with confidence. Shakedown is the initial redistribution of stresses induced by plasticity during the first loading of a structure. It relieves high stresses at corners, introduced by design, or high stresses at defects, introduced by manufacture. Contrast this with a strong ceramic material in which high stresses at a notch or flaw are always available to drive fracture because there are no mechanisms for relieving stresses. Ceramic materials are strong, hard, and abrasion-, corrosion- and temperature-resistant, but they find no place in the general world of load-bearing structures because they don't shake down. Their list of desirable properties guarantees periodic research efforts into finding toughening mechanisms for ceramic materials.

Understanding the concept of plasticity allows an appreciation of the benefits of proof loading. Many safety-critical structures are required to be loaded to well above their design load in a proof test before use. Designers gain confidence in the calculated factor of safety of a specific manufactured structure. If the structure survives, when the load is reduced, residual compressive stresses remain to protect highly loaded areas. Again, by way of contrast, a ceramic structure is likely to be made worse by proof loading because, even if the structure survives, sub-critical cracks may grow during test, leaving larger flaws behind.

In the 1930s Heinz Neuber, working on sharp notches, modelled a notch with a plastic zone as equivalent to a notch with a larger radius: the effective radius of a sharp notch being proportional to the plastic zone size at its tip. This is a routine engineering mechanics 'trick' for using an elastic analysis, in conjunction with a materials property such as the yield stress, to scale the extent of plasticity. Neuber's definitive compendium of stress concentration factors, published in 1946, provided an important input to the fatigue community until the advent of numerical solutions.

Before the advent of numerical algorithms, and computers to run them, routine analysis in three dimensions was not possible. However, although multiplying a stress in one direction by a concentration factor to model a three-dimensional feature such as a notch remains useful, it was never going to provide a satisfactory picture of the complex stresses associated with a three-dimensional feature.

Triaxiality

During the first half of the twentieth century engineers realized that the simple, uniaxial stresses associated with tensile loading on a bar became complex in the presence of a notch, or when a neck formed during a tensile test. In the body of the metal, close to the change in geometry, stresses in three dimensions develop to create triaxial stress systems. It was also observed that triaxial stress systems could inhibit ductile flow and so affect the 'yield strength' or the 'cohesive strength' of the metal under test.

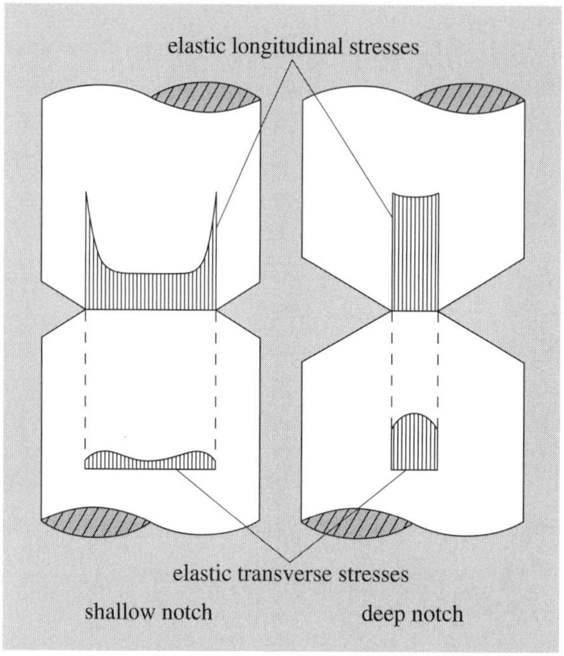

Figure 21 Stress distribution according to Sachs (upper longitudinal, lower transverse)

For example, in 1945, Sachs described the state of triaxiality associated with notches as the ratio of the average of the transverse stress on the section to the average of the longitudinal stresses (Figure 21). He also compared this with a measure based on the difference in strength between a smooth and a notched bar. In one case he had a stress measure, and in the other a failure-based criterion.

Modern finite element analysis commonly uses the following general description of the degree of triaxiality, *t*, which can be calculated automatically within the code:

$$t = \frac{\sigma_{mean}}{\sigma_{effective}}$$

where $$\sigma_{mean} = \frac{\sigma_{xx} + \sigma_{yy} + \sigma_{zz}}{3}$$

$$\sigma_{effective} = \frac{1}{\sqrt{2}}\sqrt{(\sigma_{xx} - \sigma_{yy})^2 + (\sigma_{yy} - \sigma_{zz})^2 + (\sigma_{zz} - \sigma_{xx})^2}$$

The effective stress is, in essence, equivalent to a combination of the shear stresses in all directions made directionless by squaring the terms. So this measure of triaxiality is a ratio of normal stresses in all directions to shear stresses in all directions.

Putting σ_{xx} and σ_{zz} equal to 0 in the equation above, simulating a tensile test, gives a degree of triaxiality of 0.33, which is remarkably close to Sachs's measures of 0.31 and 0.33. Producing such a single number to describe a constraint is a useful measure with which to compare different conditions. Finite-element calculations on notched round bars produce triaxiality values in the range 0.6–2.2. A lower bound value of triaxiality of 1.94 and an upper bound of 3.3 at a crack tip come from theory using typical metal properties. The greater the value of t, the more ductility is inhibited.

Thinking about a crack front that extends from one free edge to another through the body of a test piece leads to the notion that triaxiality must vary along a crack front. The stresses and strains at the free edges will be different from those deep in the metal, at the centre of the crack front. These extreme conditions can be usefully visualized as those that describe stresses in thin sheets (i.e. at the surface) and strains in thick plates (i.e. deep in the material) respectively.

The general equations for linear elastic stress and strain in three dimensions are:

$$\varepsilon_{xx} = \frac{\sigma_{xx}}{E} - \frac{\nu}{E}(\sigma_{yy} + \sigma_{zz})$$

$$\varepsilon_{yy} = \frac{\sigma_{yy}}{E} - \frac{\nu}{E}(\sigma_{xx} + \sigma_{zz})$$

$$\varepsilon_{zz} = \frac{\sigma_{zz}}{E} - \frac{\nu}{E}(\sigma_{xx} + \sigma_{yy})$$

They show how strain and stress in the same direction are linearly related through E, the Elastic or Young's modulus, and how stresses in the orthogonal directions vary the strain by Poisson's Ratio, ν.

Modelling a very thin plate is simplified by assuming that the z stresses out of the plane of the sheet are zero, as they must be at the two boundaries. This causes the general equations to reduce to:

$$\varepsilon_{xx} = \frac{\sigma_{xx}}{E} - \frac{v}{E}\left(\sigma_{yy}\right)$$

$$\varepsilon_{yy} = \frac{\sigma_{yy}}{E} - \frac{v}{E}\left(\sigma_{xx}\right)$$

$$\varepsilon_{zz} = -\frac{v}{E}\left(\sigma_{xx} + \sigma_{yy}\right)$$

These are the equations of plane stress: σ_{zz} equals zero because the sheet is free to contract in the z direction, and so ε_{zz}, the 'thinning' strain, is governed by the in-plane stretching *via* the Poisson effect.

Modelling a very thick sheet is simplified by taking the strain in the z direction as zero because contraction is inhibited in the z direction, in which case the general equations reduce to:

$$\varepsilon_{xx} = \frac{\sigma_{xx}}{E} - \frac{v}{E}\left(\sigma_{yy} + \sigma_{zz}\right)$$

$$\varepsilon_{yy} = \frac{\sigma_{yy}}{E} - \frac{v}{E}\left(\sigma_{xx} + \sigma_{zz}\right)$$

$$\varepsilon_{zz} = \frac{\sigma_{zz}}{E} - \frac{v}{E}\left(\sigma_{xx} + \sigma_{yy}\right) = 0$$

So

$$\sigma_{zz} = v\left(\sigma_{xx} + \sigma_{yy}\right)$$

These are the equations of plane strain: ε_{zz} is zero because of constraint in the z direction, which requires an out-of-plane stress, σ_{zz}.

In a thick plate containing a crack, the crack-tip stresses along the crack front can be visualized as changing from plane stress at the two free edges towards plane strain at its centre.

Further investigation of these stress conditions also shows that the direction of maximum shear stresses changes from being in the plane of the plate at its centre to being out of the plane of the plate at its edges. The effect of these changes in stress along the tip of a crack results in a commensurate change in shape of its plastic zone.

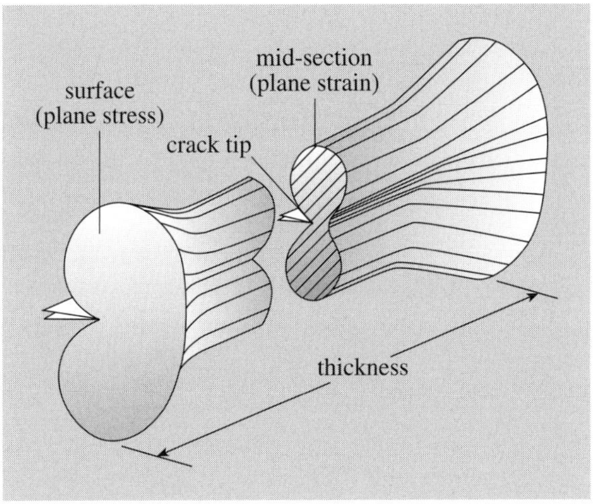

Figure 22 The change in shape of a plastic zone along a crack front

Figure 22 shows the change in shape of a plastic zone ahead of the crack tip. The loss of constraint, or decreasing triaxiality, as the free edges are approached alters the shape and size of the plastic zone that develops along a crack front. This potentially alters the way in which a metal deforms, cracks or tears at the edges when compared with the centre. A simple model estimates the plastic zone at the edges to be about three times the size of that in the centre of the crack.

A growing crack that produces dimpled rupture on the fracture surface drives its plastic zone ahead of the crack tip through a metal. It is the need to do this plastic work that confers toughness on a metal and enables it to resist crack growth.

The bigger plastic zones at the edges of a crack front will absorb more energy than the smaller zone at the centre, so it is plane strain conditions that generate the low-energy fractures associated with potentially dangerous, brittle fracture.

The change from plane strain to plane stress along a crack front demands that the effect of size on toughness is addressed. This allows the performance of a simple test piece to be related to the performance of a complex structure or part.

Size matters

Between 1920 and 1930 the assumption of scalability in toughness tests was destroyed by experiments at the National Physical Laboratory (NPL) in Teddington, UK. It is easy to underestimate the importance of this work. It showed that increasing the size of a test bar reduced the work of fracture. A tensile test on a 1-inch and 2-inch bar produced the same result – but this might not be true for a toughness test. It required publication of careful work under static loadings before the community of practice accepted that results from cracks and sharp notches in test pieces used for measuring toughness could not be scaled like those from bars in tensile tests used to measure yielding.

However, it is important to recognize that the engineering ethos at the time was concerned with understanding and modelling manufactured shapes. Large amounts of data were generated on those shapes of interest, such as gear teeth and screw threads. In designers' minds there were no sharp cracks, such as those generated during fabrication by welding that we now accept as routine. On the whole, designers avoid very sharp notches.

For large welded structures the Charpy test 'runs out of steam' because sections in pressure vessels are very much bigger than the sections tested by a swinging hammer. The inevitable welding cracks are much sharper than designed notches due to corners or intersections.

As ever, the driving force for developing theory and understanding is the occurrence of failure in a safety-critical industry. In 1965 a 167-tonne steel vessel, designed for use as an ammonia cracker, failed under hydraulic test at a (low) pressure of 34 MN m^{-2}. The design working pressure of the 18 m long, 1.7 m diameter, 150 mm wall-thickness vessel was 35 MN m^{-2}. The proof test was to be performed at 48 MN m^{-2}, as specified in the pressure-vessel codes.

After 'a kind of dull thud' the welded forging broke into four large pieces, one of which went through the workshop wall and deposited its 2 tonnes neatly in the car park, Figure 23.

The crack that caused the accident grew from a small (5 mm) embedded welding crack in the heat-affected zone of the submerged arc weld, Figure 24. There were certainly some residual stresses present as these are terribly difficult to remove completely by post-weld heat treatment. The Charpy V notch energies of the weld metal at the test temperature were roughly half of the specified value: the material was on its lower shelf.

Figure 23 Oops!

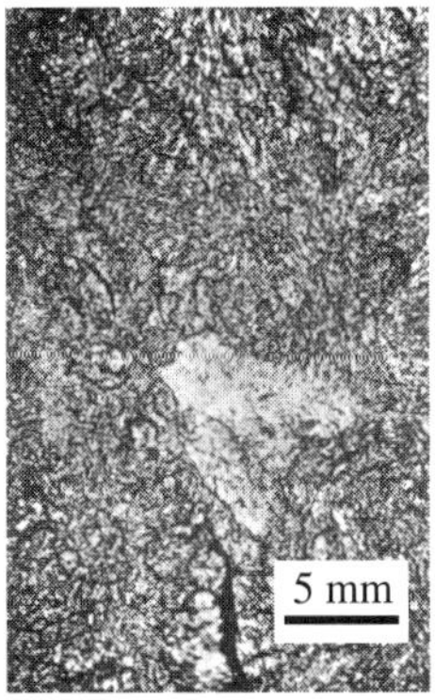

Figure 24 The initial flaw

The British Welding Research Association (later to become The Welding Institute and currently called TWI) recommended that 'fracture mechanics principles be used when setting fracture avoidance criteria for thick high-strength steels'. The manufacturer had the pieces welded back together and delivered the vessel to its customer, who is using it still.

Although the vessel described above is large by the standards of the petrochemical industry for which it was made, it is small by comparison with a nuclear pressure vessel. The consequences of failure in a nuclear installation are much more severe. A nuclear vessel falls neatly into Irwin's category of an uninsurable risk as a driving force for technical development.

The industrial drivers changed with the advent of nuclear power plant. Not only had the size of manufactured structures left the small Charpy toughness specimens far behind, but there was an urgent need for predictive capability from test-piece measurements that the Charpy test could not deliver.

So, a new test was required to measure the minimum energy fractures associated with plane strain conditions. I have developed enough physical principles at this stage in the narrative to explain the design of such a test.

Designing a toughness test

To cope with failures such as the ammonia vessel, and potentially more dreadful failures, required engineers to embrace the problem of very sharp cracks in thick sections from theoretical, experimental and practical standpoints.

From an experimental viewpoint the obvious first step is to start with the sharpest crack that can practically be placed into the body of a test piece. This is a fatigue crack grown at a low load range.

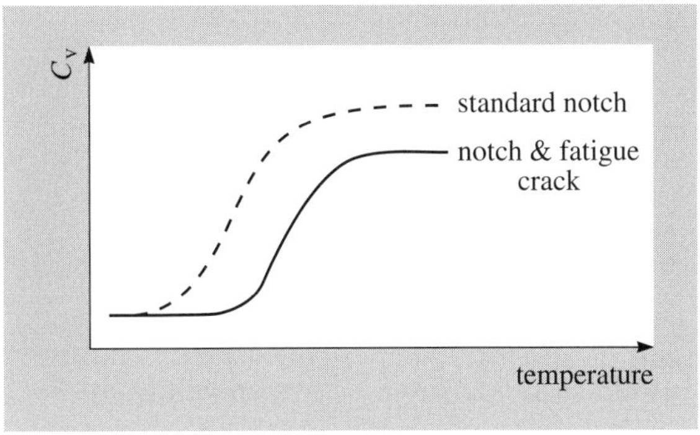

Figure 25 Sharpening a Charpy notch (cf. Figure 17)

Putting a crack into a standard Charpy test piece by fatiguing the specimen from its notch vividly illustrates the effect, producing a drop in toughness of about 20% (Figure 25). The crack was grown well into the body of the section, away from the effect of plasticity around the notch.

The general availability of hydraulic testing machines means that slow tensile tests have replaced the uncertainties and difficulties associated with the use of swinging hammers and flying halves of test pieces. There is the small complication of putting similar fatigue cracks into every specimen tested; special machines are sold purely for the purpose of growing fatigue cracks, from notches, under controlled conditions. So, all of those earlier arguments about the ease and cheapness of a toughness test have been challenged.

The next major problem to deal with is that 'similitude' (scalability) does not apply to toughness tests. It is now accepted that a material will produce different values of toughness when tested at different sizes. Therefore the sensible aim is to produce a measure of toughness that is independent of size and so represents a genuine materials property. Before pursuing this issue further, let us look at some size effects in different materials in order to illustrate and emphasize the extent of the problem.

Fracture surfaces

Figure 26 Ductile tearing

The fracture surface shown in Figure 26 comes from a massive test specimen, some 14 cm wide, made from a high-quality, carbon–manganese, or 'mild' steel.

The first marking at the lower edge is from the machined notch used to start a fatigue crack. The second, smooth feature, with a curved front boundary, is from the fatigue crack that was grown into the body of the steel, prior to fracturing by overload. The main fracture surface shows plastic deformation at the sides and massive, energy-consuming, ductile tearing, despite the constraint of the section size.

The similar-sized specimen shown in Figure 27 illustrates how plane strain constraint in the body of the steel has caused secondary cracking due to the stresses set up parallel to the crack border. The 'cheeks' of the crack are free surfaces capable of deforming plastically, but strain is constrained in the body of the specimen, so allowing the stresses to develop that tore the material apart.

If you are wondering why these pictures are 'slivers' of the fracture surfaces, rather than complete specimens, it is because I would have needed a wheelbarrow to carry them around! By contrast I can, and do, carry the 15 mm wide specimen in Figure 28 around in my briefcase. It is a high strength, hard steel that has fractured in a perfectly planar fashion. It shows the 'sparkling' appearance typical of a low-energy fracture by cleavage of the metal grains. Comparing the two different types of surface it is easy to see why a nineteenth-century observer might be betrayed into thinking that brittle fractures were caused by the state of a material changing in some way: by 'becoming crystalline'.

Figure 27 Transverse cracking due to constraint

Figure 28 A classic brittle fracture

The two specimens considered so far are classic representatives of extremes of fracture behaviour, so their load–deflection or stress–strain traces are also classic examples of ductile and brittle fractures. The trace for a brittle failure is simply a straight line to fracture and the ductile fracture trace becomes non-linear after an initial, elastic response. So the amount of non-linearity is a sensible way to classify a fracture, in addition to the evidence from the appearance of its surface.

Figure 29 Shear lips

The specimen shown in Figure 29 is about the size of my thumb, so I can easily carry it around in my pocket. It is a high-strength nickel alloy of a type used in jet engines.

The machined notch, the smooth surface of the fatigue crack and the transition from shear lips at the edges of the crack to a dull planar central section are clearly shown. The shear lips are evidence of the change from plane stress at the edges of the specimen, to plane strain over the planar section. That the fracture surface is dull suggests that there is dimpled rupture over the surface.

To make sense of the effect of specimen size, engineers measured the toughness of test pieces made from the same material at different thicknesses. The toughness measurements reduced asymptotically to a minimum value as the width of the specimen increased, Figure 30.

When the shear lips are only a small proportion of the fracture surface the planar section dominates. Therefore the fracture toughness is a minimum that can be taken as a materials property independent of specimen size. It is important to recognize that the planar surface might show cleavage, dimpled rupture or a mixture of both at the microstructural level.

This principle applies to all metals. However, some metals require large sizes that are impossible to test, whereas some produce brittle fractures in small, practical sizes. In principle it would be possible to produce planar fractures in mild steel, but the specimens would need to be the size of a filing cabinet. So, we could not build the machines to break them, nor the rolling mills to make them!

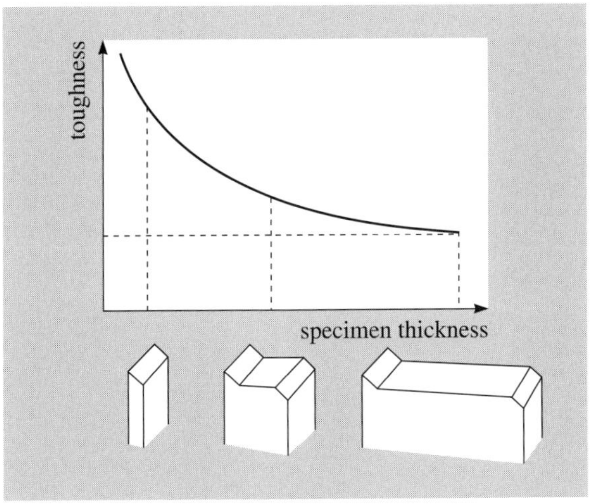

Figure 30 Less lip, more toughness

The problem for designers of standards is to frame a set of rules that produce a valid measure of fracture toughness for a material. These rules are enshrined in BS 7448 *Fracture mechanics toughness tests*.

As with all standards there is a compromise to be struck between necessary conservatism and practicality. So, how to allow for a specimen large enough to produce a low-energy fracture, but not so large as to be uneconomic or impractical?

In BS 7448 two specimen forms are allowed; the compact tension specimen (CT), shown in Figure 31, and the three-point bend specimen (TPB) shown in Figure 32.

The CT specimen fulfils the earlier dictum of reducing the amount of metal needed for a test because the arms of a TPB specimen are only there to transport load to the critical section. However, there are more subtle differences between the two test geometries. The first is that there are different traditions of testing in different industries. TWI will tend to use the more traditional bend tests because its expertise grew out of welding and testing high-toughness, low-strength structural steels. The way the length of a beam magnifies load is extremely useful because bigger cross-sections can be tested without the need for a more powerful machine. The aircraft industry, which uses high-strength, low-toughness materials, uses the newer CT geometry almost exclusively.

In principle, a CT test will give a more conservative result because of the intensity of opening stresses ahead of the crack tip. The far side of the ligament ahead of a crack in a bend specimen will always contain compressive stresses whereas the CT specimen combines bending with opening stresses that negate the compressive stress field.

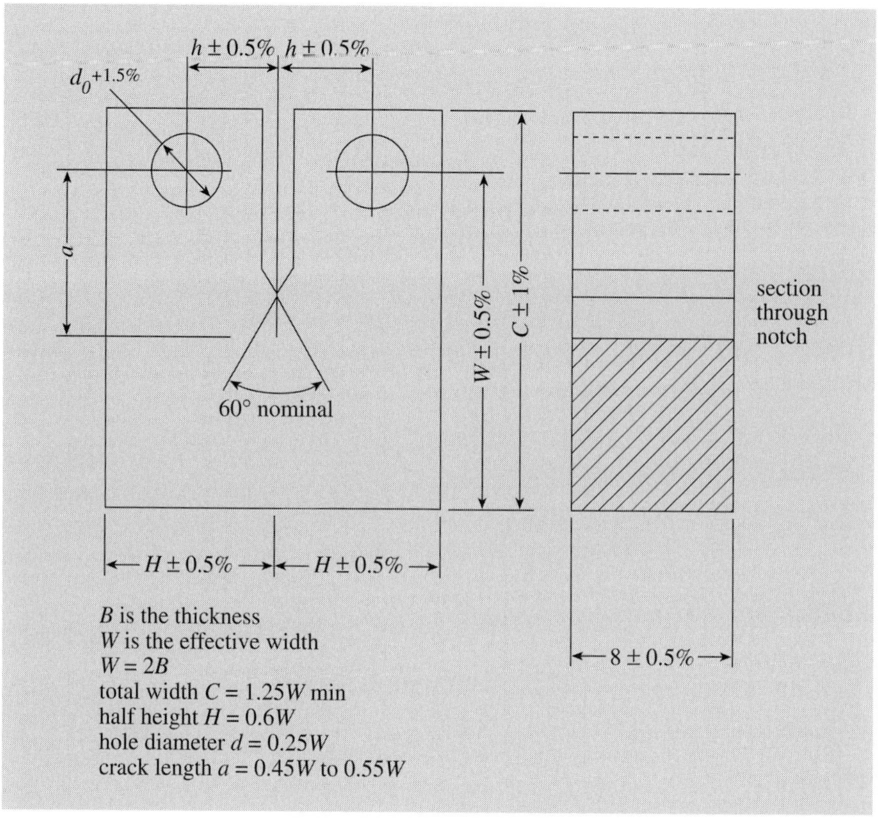

Figure 31 Compact tension specimen

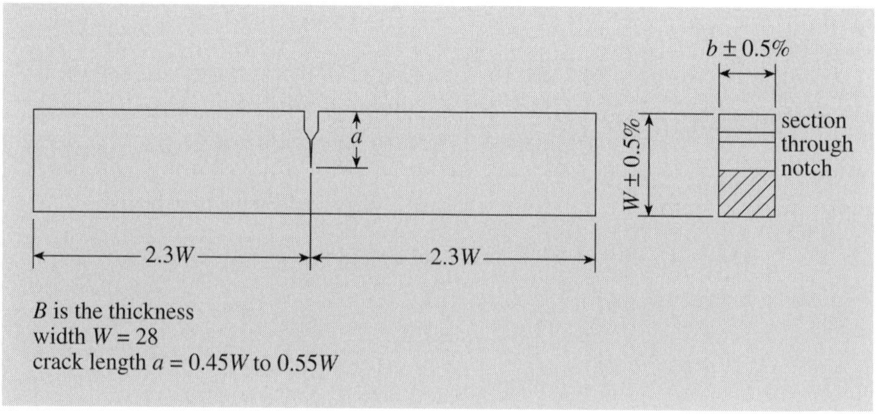

Figure 32 Three-point bend specimen

After a wealth of detail on test-piece shapes and how to get a sharp fatigue crack into the body of a test-piece, the heart of the standard lies in the interpretation of the load–deflection trace, for it is hard to frame a set of rules based on the pictorial interpretation of a fractured surface.

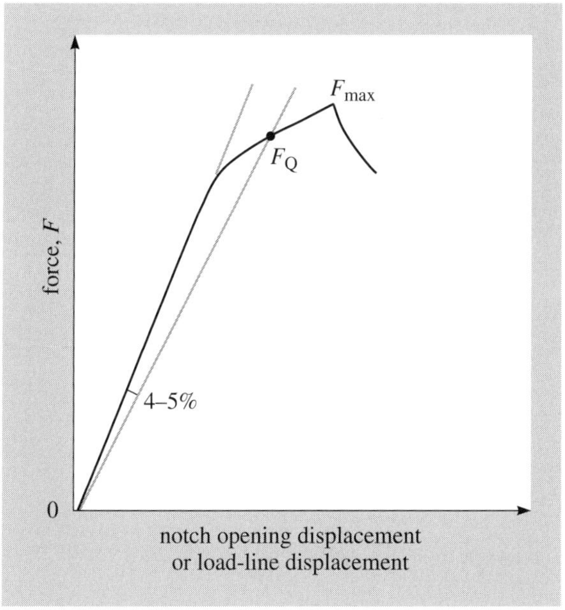

Figure 33 Interpreting non-linearity

The amount of non-linearity allowed in a load–deflection trace depends on the shape of the specimen tested and the way in which its displacement is measured. From experience, it is usually specified from a line with a slope 4% or 5% less than a tangent to the curve. Figure 33 shows one example. A load value on the trace, F_Q, is determined and used to calculate the value of toughness. However, this value of toughness is only acceptable as a materials property if the calculated size of its plastic zone at that load is less than about a fiftieth of the bulk of metal encapsulating the zone. So, in principle, a tester does not know the size of specimen required before performing a test. Standards offer some guidance, but it is not until after a test piece has fractured that the validity of a test can be determined. It may be a good idea to use an experienced test house.

There is a continual, international debate about the conservatism of rules for testing. It is possible to fracture a steel test piece that shows a classical planar fracture surface, but which fails to meet the conditions of the standard. Standards develop over time, and different industries and test-houses develop their own expertise with different types of metals and geometries.

The specimen in Figure 34 shows the use of side grooves to add extra constraint, and so inhibit ductility at the sides of a crack where conditions approach those of plane stress. These side grooves enable low-energy fractures to be obtained in smaller specimens.

Figure 34 Added constraint from side grooves

Growing a fatigue crack with a straight front for testing a pressure vessel steel is difficult enough, but some very high-strength aerospace alloys pose even trickier problems.

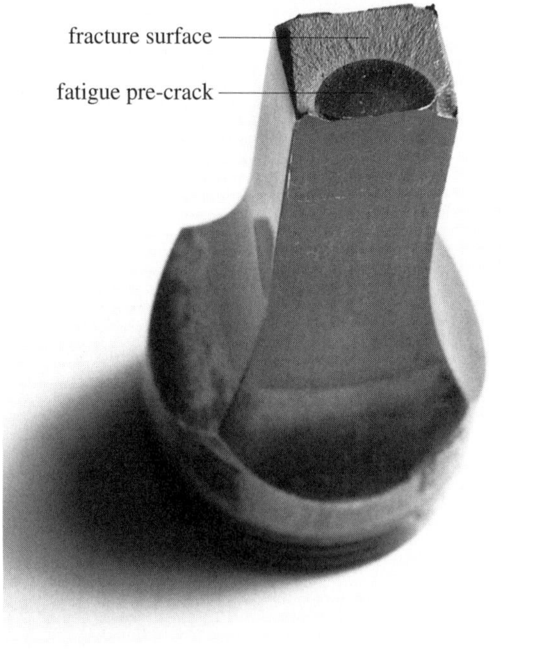

Figure 35 Heat tinting a fatigue crack

Figure 35 shows a fatigue crack grown deep into the body of a nickel alloy test piece. Note the care with which the test piece was designed to transmit load smoothly into the small test section from the large, screwed ends. The fatigue crack was started in a fin that was ground off after the crack had grown radially into the section. The fatigue crack was then heat-tinted before the test-piece was broken, so that the extent of the pre-crack could be easily measured.

Such tests are specific to the industries that develop them. They have no place in the international standards which allow measurements to be made that can be published, exchanged, bought and sold with confidence.

So to summarize, in the fracture mechanics toughness tests just described the geometry variable has been taken out of the Charpy test in order to produce a materials property that describes the minimum effort to fracture a material: a genuine materials property.

There has, nevertheless, been something of a sleight of hand in the development of the story. The calculation of a toughness value from a load measurement is accomplished by using the geometry of the specimen alone, so a modern toughness value must come from a stress-based theory, not an energy-based theory. Stress is the 'bread and butter' of predictive mechanics; moving from measuring toughness in terms of energy, as in the Charpy test, to being able to measure toughness in terms of load was a godsend to engineers.

How this change in culture happened is a classic example of a paradigm change in the development of ideas on fracture and failure. To understand this, we must consider the pioneering work of Irwin.

K is for Joe

Figure 36 George Rankine Irwin

In the 1950s George Rankine Irwin, Figure 36, a physicist working at the US Naval Research Laboratory (NRL) in Washington DC laid down the foundations of modern linear elastic fracture mechanics (LEFM). He articulated that the energy needed to extend a crack comes from the rate of loss of strain energy stored in the elastic system.

> *The governing character of the balance between work done and energy release with extension of the crack must be accepted as basic to any scheme for calculating safety against fracture propagation.*
>
> *(Irwin and Kies, 1954)*

Irwin called the strain-energy release rate *G*, after A. A. Griffith who researched the effect of flaws on the breaking of glass in the 1920s. But Irwin's ideas did not grow naturally from Griffith's; Irwin changed the way in which fracture is understood.

G was calculated from stiffness measurements on specimens containing cracks of different lengths. The difference between the areas under two stress–strain curves, representing cracks of different lengths, was taken as the elastic strain energy available to grow a crack that distance. This is a static model of a dynamic process because the two specimens are snapshots of stationary cracks.

Ballistics was the business of NRL, so a large number of experiments were performed on ship steels. The work focused on establishing whether a critical value of *G* measured in tests could be used to predict the initiation of fast fracture based on the stress and strain field around the small plastic zone at a crack tip. The critical value of *G* was called G_c, a measure of material toughness.

Up to this point Irwin was working within the established energy paradigm. Combine this research environment with the practical background of routine Charpy test measurements and it is possible to see how, from a historical perspective, the energy paradigm held sway over the fracture community at that time.

Figure 37 Joe Kies

Rossmanith in his excellent work on the history of fracture mechanics (1997) reports that US aircraft engineers coined the use of K as a measure of toughness. They wrote fracture results as $\sqrt{G_c E}$, which they called K_c, after Joe Kies, Figure 37. It was he who had pointed out that the critical stress for a given crack size depended on this value. Kies was using the $\sqrt{G_c E}$ term, computed only from the applied stress and the crack size, for a test to describe fracture properties of hot-stretched acrylics, a material now in routine use for aircraft windows.

As strain energy density is related to the product $\sigma\varepsilon$, and ε is proportional to σ/E, then strain energy density will be related to σ^2/E. So, multiplying an energy term by a modulus will produce a stress squared term. However, one would not necessarily expect to be able to 'remove' the strain term altogether and so produce a one-parameter approach using only the stress field around the small plastic zone at a crack tip, but that is what happened.

Linear elastic fracture mechanics (LEFM)

It took a long time, but the fundamental importance of the form of the stress distribution ahead of a crack in a large plate subject to an opening load has now been recognized. H. M. Westergaard, working on bearing contact pressures in 1939, developed elastic solution techniques that were seized upon by Irwin to characterize crack-tip stress fields.

The result, which looks terrifying, is:

$$\sigma_{xx} = \left(\frac{K_I}{(2\pi r)^{0.5}}\right)\left(\cos\frac{\theta}{2}\left[1 - \sin\frac{\theta}{2}\sin\frac{3\theta}{2}\right]\right)$$

$$\sigma_{yy} = \left(\frac{K_I}{(2\pi r)^{0.5}}\right)\left(\cos\frac{\theta}{2}\left[1 + \sin\frac{\theta}{2}\sin\frac{3\theta}{2}\right]\right)$$

$$\sigma_{xy} = \left(\frac{K_I}{(2\pi r)^{0.5}}\right)\left(\cos\frac{\theta}{2}\left[\sin\frac{\theta}{2}\cos\frac{3\theta}{2}\right]\right)$$

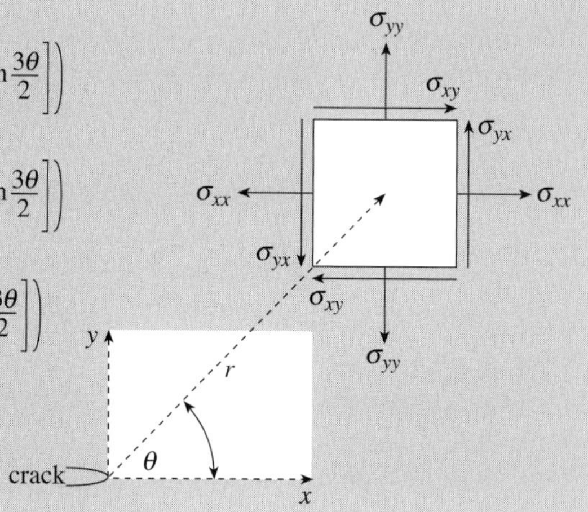

The key point to grasp is the form of the stress field ahead of a crack. Figure 38 shows the shape of the trigonometric functions in the second pair of brackets for each of the three equations as θ changes from −180° to +180°.

The subscript in K_I indicates that the primary loads are opening the crack faces. Other subscripts (K_{II} and K_{III}) are used for shear loading of the crack faces. In homogeneous metals, cracks generally grow in the direction of the largest opening loads so other forms of loading will not be pursued. If the subscript is missing it is generally safe to assume that it is the opening mode that is being described.

Westergaard established that in a linearly elastic body containing a crack this form is always the same. So, the stress field in my test piece will have the same form as the stress field in your cracked part.

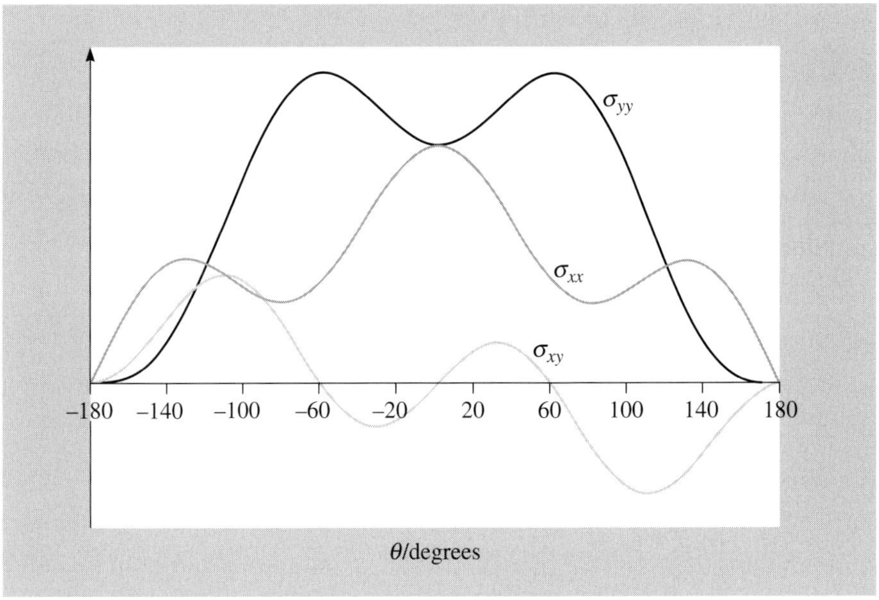

Figure 38 Characteristic forms of equations on page 52

Turning now to the terms in the first pair of brackets for each of the three equations: the stress changes in proportion to the inverse of the square root of the distance from the crack tip so, as with Inglis's solution for stress concentration factors, an infinite stress is predicted at the crack tip: a singularity.

The 2π term is not a variable, which leaves K_I as the parameter which characterizes the key, crack tip opening, stress field; it is this that governs the size of the stress field. If the crack K_I conditions are the same, regardless of the loading or generating mechanisms, you will have the same response in a component as in a test piece.

Irwin's solution can be applied in general to cracked bodies. Therefore it can be arbitrarily simplified to some geometries of interest. Removing the complication of angular variation by considering only the opening stress ahead of the crack tip gives

$$\sigma_{yy} = \frac{K_I}{\sqrt{2\pi x}}$$

where x = distance ahead of the crack

There are only two pieces of information needed to fully describe a centre crack in an infinite plate: the remote stress, σ, and the length of crack, $2a$ by convention, so we can write

$$K_I \approx \sigma\sqrt{\pi a}$$

and deduce that its units are MN $m^{-3/2}$, or MPa $\sqrt{m}$.

Putting in a dimensionless scaling factor, Y, to produce an equation gives

$$K_{\mathrm{I}} = Y\sigma\sqrt{\pi a}$$

Unfortunately, history dictates that π remains inside the square root.

The value of Y accounts for different geometrical and loading effects, such as the nearness of approach of the crack tip to a free surface, the curvature of a free surface or a component, and the type of loading.

For different geometries and types of loading values for Y are found from handbooks of numerical solutions. These usually contain a single curve-fitting expression created from a number of individual numerical solutions. For fracture mechanics practitioners using proprietary software these factors are contained in their computer codes.

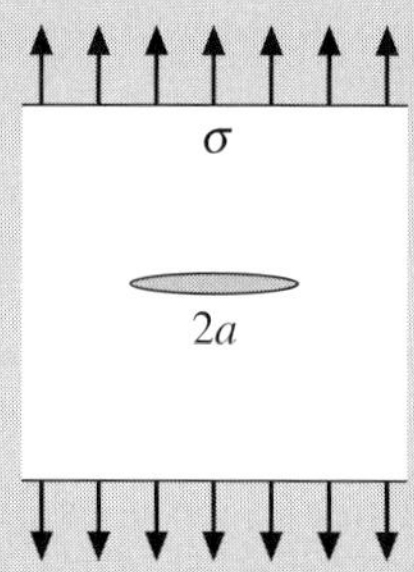

For a centre cracked infinite plate Y = 1.
Inserting Y = 1 into

$$K_1 = Y\sigma\sqrt{\pi a}$$

gives

$$K_1 = 1\sigma\sqrt{\pi a} = \sigma\sqrt{\pi a}$$

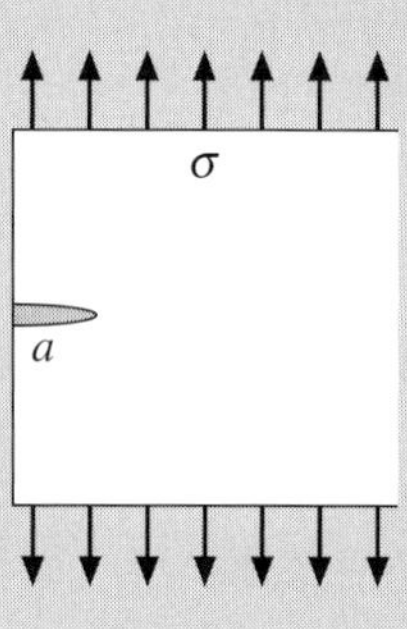

For an edge cracked infinite plate $Y = 1.12$. Note that for an edge crack its length is a, whereas it is $2a$ for a centre crack: ‘two tips $2a$’.

$$K_1 = Y\sigma\sqrt{\pi a}$$

$$K_1 = 1.12\sigma\sqrt{\pi a}$$

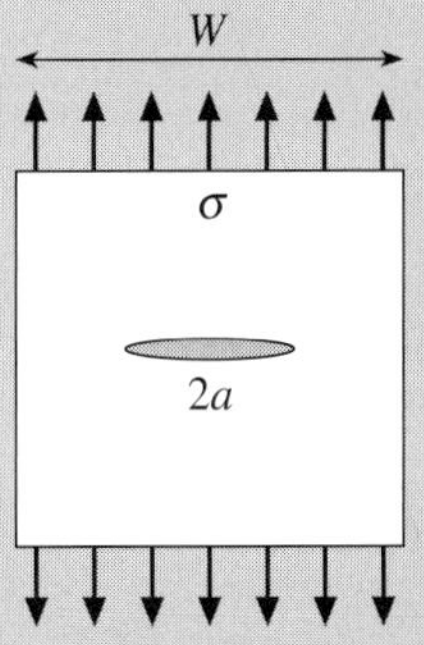

For an embedded crack in a plate of finite width, the crack tip stresses are affected by the free surface ahead of the crack. Therefore the value of Y must be a function of this geometry effect, described in terms of a/W, in order to keep Y dimensionless. The value of Y increases as the crack approaches the free surfaces.

$$K_1 = Y\sigma\sqrt{\pi a}$$

$$K_1 = \left(\frac{1 - 0.025\left(\frac{a}{W}\right)^2 + 0.06\left(\frac{a}{W}\right)^4}{\sqrt{\cos\left(\frac{\pi a}{2W}\right)}} \right) \sigma\sqrt{\pi a}$$

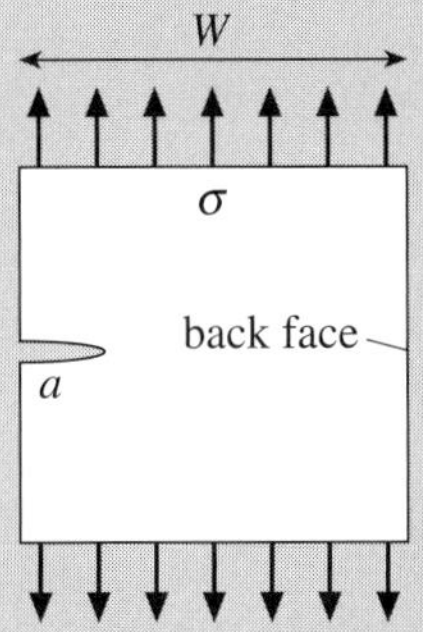

Similarly the stress field ahead of an edge crack is affected by its proximity to the back face, again as a function of a/W to preserve the dimensionless nature of Y. The value of Y increases as the crack approaches the back face.

$$K_1 = Y\sigma\sqrt{\pi a}$$

$$K_1 = \left(1.12 - 0.23\frac{a}{W} + 10.55\left(\frac{a}{W}\right)^2 - 21.72\left(\frac{a}{W}\right)^3 + 30.39\left(\frac{a}{W}\right)^4\right)\sigma\sqrt{\pi a}$$

Values of K_I can be added together for combined loadings of, say, tension and bending if Y is known separately for each case. Do not add stresses! K_I is a mechanics parameter so, as with stress, it can take any value that the load and geometry dictate until a material event is reached (for example, in a tensile test, it is yield). In the case of LEFM it is fast, brittle fracture that intervenes.

Matching a mechanics parameter to a materials parameter, as in $\sigma = \sigma_Y$, is the natural way for engineers to express a failure criterion. So, $K_I = K_{IC}$ becomes the way to express a brittle fracture criterion, with K_{IC} as the fracture toughness of a material. K_{IC} is the property measured in a fracture toughness test conducted according to BS 7448.

Irwin's LEFM theory also provides us with the means for a rough and ready estimate of the size of a plastic zone at a crack tip. The radius of a notionally circular plastic zone, r_p, is given by:

$$r_p = \frac{1}{3\pi}\left(\frac{K}{\sigma_Y}\right)^2 \quad \text{for plane strain}$$

$$r_p = \frac{1}{\pi}\left(\frac{K}{\sigma_Y}\right)^2 \quad \text{for plane stress}$$

The values of 3π and π are not particularly special; there are better estimates in the literature.

The size relationship between a plastic zone and its surrounding elastic material is used to validate fracture toughness tests. All standards have a size specified as some fraction of

$$\left(\frac{K_I}{\sigma_Y}\right)^2$$

in order to ensure that a test is valid.

Small-scale yielding is the term used to describe a tiny plastic zone surrounded by a large bulk of elastic material. Under conditions of small-scale yielding it is the *K* field that dominates the fracture event, not the small amounts of yielding at the crack tip.

K_I could be related to the previous energy-based approach through $K_I^2 = GE$ and consequently at fracture, $K_{IC}^2 = G_cE$. Irwin needed K_I, the single, crack-tip characterizing parameter, to make a stress-based approach to fracture possible.

However, gaining acceptance for a new material property was never going to be easy. What was needed was a body of evidence that supported *K* as a useful crack-tip characterizing parameter that made the prediction of fracture in a large structure from tests on a small specimen possible. This process took the best part of a quarter of a century.

Fracture toughness values

Fracture mechanics ideas based upon stress have been applied to all classes of materials over the years. Considerable expertise has accrued in industry, supported by government laboratories and universities. In the table below are approximate values of K_{IC} for useful, structural metals, together with glass which exhibits the closest to perfectly brittle fracture of an everyday material.

Material	K_{IC} value (MN $m^{-3/2}$)
High-strength, modern steels	50–200
Older, quality steels	Down to 40
Titanium alloys	50–100
Aluminium alloys	20–50
Cast irons	6–20
Glass	0.5

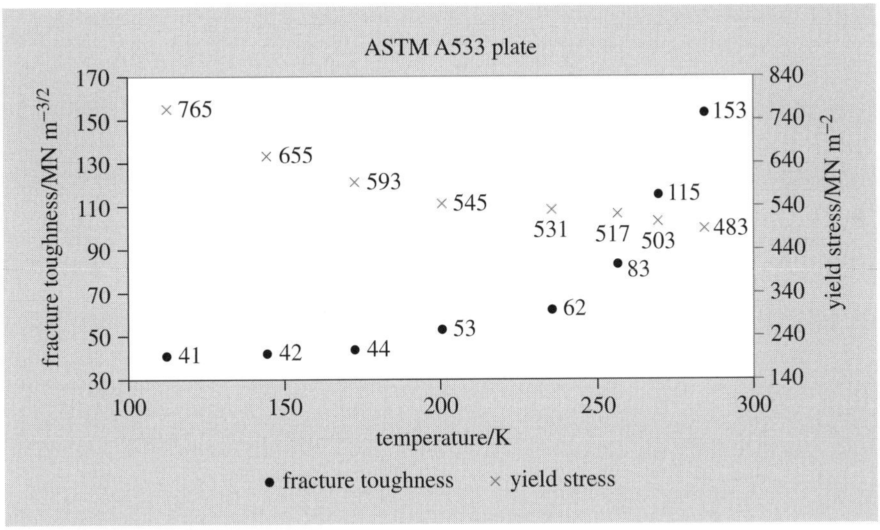

Figure 39 Toughness and yield change with temperature

The vitally important temperature dependence of materials properties in ferritic steels is vividly shown in Figure 39. The yield stress of this 0.24C steel increases by 58% and the fracture toughness falls by 73% over the same temperature range. Plate a foot thick was tested in order to produce the fracture toughness data point shown at the highest temperature on the graph.

Buying a value of fracture toughness to British Standards is expensive, about £250 a number at the start of 2002. The cost of producing Figure 39 was borne by the US Government. By contrast a Charpy test is about a tenth of the price.

Clearly, buying a lot of fracture toughness measurements for quality control or routine use is prohibitively expensive. There is a constant interest in the possibility of relating Charpy values to K_{IC} values.

All of the principles so far discussed militate against the likelihood of a useful general relationship between Charpy toughness and K_{IC}. In particular, Charpy values of a steel on its upper shelf involve the initiation and growth of a tough fracture. Nevertheless, many attempts have been made to develop relationships for particular materials, or to draw lower bound curves under masses of data to produce conservative (often very conservative) values. Modern-pressure vessel steels with Charpy Energy (C_V) values no greater than 70 Joules or less than 7 Joules were studied in 1973 to produce

$$K_{IC} = 14.7{C_V}^{1/2}$$

with a scatter band of ±15%.

A recent Structural Integrity Assessment Procedure for European Industry (SINTAP) (Webster and Bannister, 1999) document suggests a lower bound relationship for steel of

$$K_{IC} = 12{C_V}^{1/2}$$

These relationships need to be used with great care because they are entirely empirical. None of these relationships works for older, less clean steels that can produce fracture toughness values less than 40 'fracture mechanics units'.

Calculating with *K*

Matching a $K_{\text{crack driving force}}$ to a $K_{\text{material property}}$ gives us

$$Y\sigma\sqrt{\pi a} = K_{IC}$$

The value of *Y* comes from a handbook, or is bought. The remote stress field, σ, is calculated from a component's load and geometry. The crack length, a, is measured or estimated and K_{IC} is measured, or taken from others' measurements on the same material, in the same heat-treated state and at the same temperature.

From knowing any two of the engineering parameters (a, σ and K_{IC}) the third can be calculated. For example, knowing the material from which a component is made and measuring the length of a pre-crack that caused a catastrophic failure allows the stress and hence the load at which failure occurred to be calculated. This is a huge asset to a failure investigator as the calculation comes directly from the failure site. Taking a measured crack length in a component under its design load, made from a well-characterized material, allows a plant or aircraft operator to decide whether the component can stay in service, and how long the crack needs to be in order to initiate catastrophic failure.

So, all of the historical work on LEFM leads to a remarkably simple technique with which to perform lower-bound, or safe, calculations and predictions. The support of numerical analysis and the accumulation of a body of materials data for important engineering metals have made this possible. Brittle fracture is the form of failure that needs the least energy to drive a crack. To improve predictive capabilities beyond a lower-bound calculation requires theories that are relevant to tougher fractures. Before continuing with the main narrative on overload failure I shall take a short detour into the world of sub-critical crack growth by fatigue.

Fatigue

It is possible for cracks to grow at loads well within the design envelope by a number of sub-critical crack growth mechanisms. By far the most widespread of these mechanisms is fatigue. Cyclic variation of stresses within the design envelope of a structure can drive a crack forward by ratcheting it through its plastic zone.

With hindsight the application of LEFM to fatigue seems to be an obvious extension of the ideas coming from brittle-fracture research, because a fatigue crack growing slowly and stably in the body of a metal component has a small plastic zone surrounded by bulk elastic material.

Figure 40 Paul Paris

It was Paul Paris, Figure 40, working at Boeing in 1957, who proposed the application of LEFM theory to fatigue. Prompted by Irwin's crack-tip stress field analysis, Paris's insight was not easy to bring to fruition. There were no data for Paris to work with until 1959; despite the routine observation of cracks in all monocoque airframes and the Comet crash at the beginning of the decade. Paris's view on this was unequivocal:

> *Although I have since obtained a commercial pilots licence, it is only honest and informative to note that from 1954 until 1965, I declined to travel by air even on trips for the Boeing Company.*

Paris put together data from three independent sources to produce a compelling case for the correlation between *K* and fatigue crack growth. His paper was rejected by three journals whose reviewers opined

> *it is not possible that an elastic parameter such as K can account for the self-evident plasticity effects in correlating fatigue crack growth rates.*

Materials scientists in the 1950s were very resistant to the ideas of fracture mechanics, viewing approaches that were not based on dislocation theory as being fundamentally unsound. The argument of these reviewers could also be used against LEFM theory applied to stationary cracks! Paris's seminal work was circulated inside Boeing, and eventually published in a minor University of Washington periodical in 1961 (Paris et al., 1961).

Throughout the 1960s there was little industry interest until a major aircraft incident. A wing was lost on an F-111 aircraft in 1969. It was found to have been caused by the extension of a forging flaw by fatigue. The fleet was grounded for six months. Fracture mechanics solved the F-111 problem. It was politically and commercially so important that fracture mechanics moved into the vanguard of engineering techniques. Arguably, this accident introduced the notion of damage-tolerant ideas.

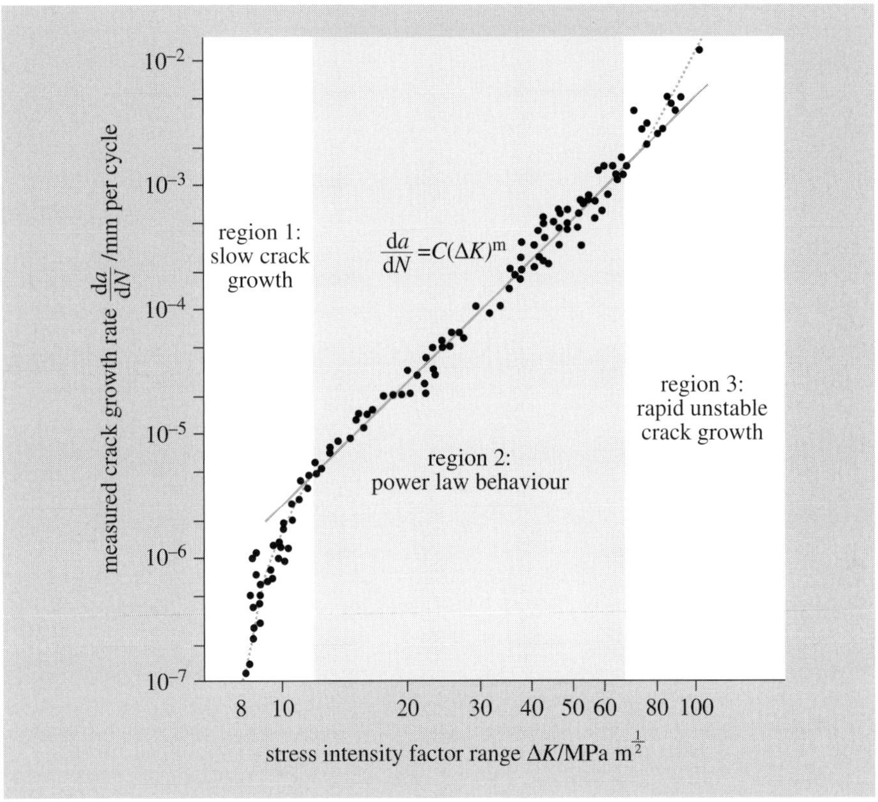

Figure 41 The Paris equation in a nutshell

The work that Paul Paris started in the late 1950s is encapsulated in Figure 41, which shows the characteristic relationship between crack growth rate and stress intensity factor range, ΔK.

Fatigue cracks are driven by the cyclical stress range ($\Delta\sigma$) to which they are exposed, not by the static stresses on a component. So, as K is related to σ, so ΔK is related to $\Delta\sigma$. The fracture mechanics equation

$$K = Y\sigma\left(\pi a\right)^{1/2}$$

becomes

$$\Delta K = Y\Delta\sigma\left(\pi a\right)^{1/2}$$

Region 2 in Figure 41 shows a power-law relationship that is broad enough to be useful for prediction and is remarkably unaffected by microstructure and mean stress effects.

The Paris equation over Region 2 is straightforward.

The curve's coefficient and exponent, C and m, can be taken as about 10^{-11} and 3 (using units of m and MN) respectively for all ferritic steels. The predictive power of this relationship is now clear and its use has been extensive.

Putting

$$\Delta K = Y\Delta\sigma\left(\pi a\right)^{1/2}$$

into the Paris equation,

$$\frac{\mathrm{d}a}{\mathrm{d}N} = C\left(\Delta K\right)^{m}$$

produces the expression

$$\frac{\mathrm{d}a}{\mathrm{d}N} = C\left(\Delta K\right)^{m} = C\left(Y\Delta\sigma\sqrt{\pi a}\right)^{m}$$

This equation can be integrated directly if Y is a constant. Numerical integration is required if Y introduces a further dependence on a that does not result in a standard integral.

In Region 1, at low values of the stress-intensity factor range, the Paris power-law relationship fails when microstructure and mean stress effects become dominant. Measurements are difficult at very low crack growth rates where a threshold value of ΔK is indicated. Less importantly, because a component has little life left in Region 3, the relationship also breaks down when K is approaching values at which fracture processes take over from fatigue processes.

Although the Paris equation has proved to be a remarkably powerful formula for calculating fatigue-crack growth in metals and metal alloys, it would be wrong to read Figure 41 as applying to cracks of all lengths. It describes the behaviour of 'long' cracks, those that are bigger than the scale of the microstructure by a factor of about four or five. Modelling the growth of short cracks has been the subject of a great deal of work in the aeroengine industry – short cracks do grow below the Paris threshold and a component typically spends most of its life containing one or more short cracks that are growing slowly.

Fatigue processes cause crack growth at stresses well below those required to initiate failure, hence the term sub-critical crack growth; but the large subject of fatigue is only a side issue here, this narrative must now return to the world of overload failure.

Plastic collapse

As the loading on a component increases it does not automatically follow that a pre-existing crack will grow, or that a crack will form from a notch. In components made from low-strength metals a crack or sharp notch might simply reduce the size of a cross-section under load. As the load on such a component increases, plasticity spreads over the whole of the reduced cross-section until it can withstand no further increase in load and so collapses.

So, this narrative continues by considering plastic collapse in order to 'bookend' the various complex fracture and failure behaviours of metals that can occur between fast, brittle fracture and plastic collapse.

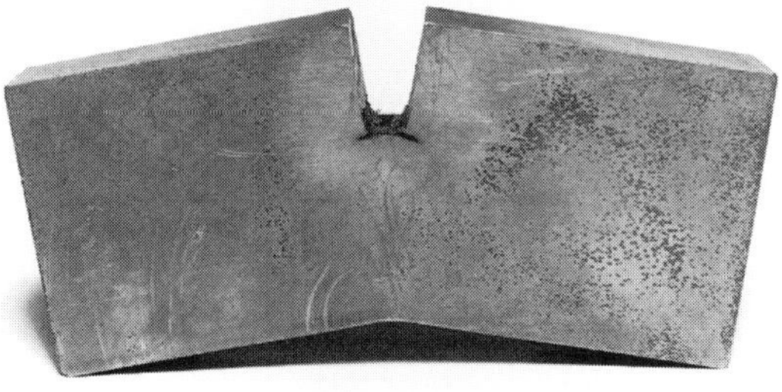

Figure 42 Fully developed plasticity

Figure 42 shows a beam in three-point bending with a notch at the position of maximum bending moment. It has been loaded until all the metal over the ligament at the reduced cross-section has been plastically deformed.

Once plasticity reaches the free surfaces of the structure it behaves as a mechanism. It is as if a hinge has formed where the notch has reduced the depth of section at the position of greatest loading. The name of this mode of failure is plastic collapse.

Plastic collapse is a relatively safe type of failure compared with fast, brittle fracture. In a beam, yielding first takes place at the extremes of the section under maximum load. Increasing the load spreads the yielded material towards the centre of the section whilst driving the already yielded material up its work-hardening curve, so requiring more and more load to progress the collapse.

Figure 43 A skeleton of rolled steel joists (RSJs)

Buildings made from grillages of rolled steel joists, Figure 43, usually offer alternative load paths through the structure. Therefore hinges will form until the structure becomes statically determinate, after which the next hinge to form creates a mechanism and the building collapses gracefully. This process is the way in which steel-framed buildings, manufactured from 'I' beams made from mild steel, are assumed to fail during a fire.

In the 1960s, before the widespread availability of numerical techniques running on digital computers, most of the analytical effort available was put into modelling plasticity for forging, extrusion and structural failure, using bounding theorems that do not require a complete stress analysis.

To solve a stress analysis problem exactly requires:

- Equilibrium to be satisfied in the elastic and plastic zones and at the free boundaries.
- Compatible strains and strain increments.
- Stresses and strains that abide by Hooke's law in the elastic region.
- That the appropriate yield criterion is satisfied in the plastic zone.

There are two bounding theorems that do not require a complete analysis: an upper bound theorem and a lower bound theorem.

An analyst chooses a deformation mechanism that allows a structure to collapse and estimates the plastic collapse load by equating the internal rate of dissipation of energy to the rate at which external forces do work on the body. The result of the analysis is either correct, or high; hence upper bound theorem.

This theorem was ideal for forging, where the worst conceivable outcome from an analysis would be that there wasn't enough load capacity in the press or draw bench to fully deform the section.

The lower bound theorem requires a stress distribution that satisfies equilibrium everywhere and does not violate the yield condition. It predicts either the correct collapse load or a load that is lower than the correct collapse load. Such an approach is ideal for structural analysis because its results are conservative.

Analysing plastic collapse

Lower bound solutions use equilibrium alone to produce a solution. In this section I show equilibrium solutions for an axial load and for a beam in bending.

Figure 44 shows a bar of rectangular cross-section BW with a crack of length a in the centre, leaving a ligament of $(W–a)$.

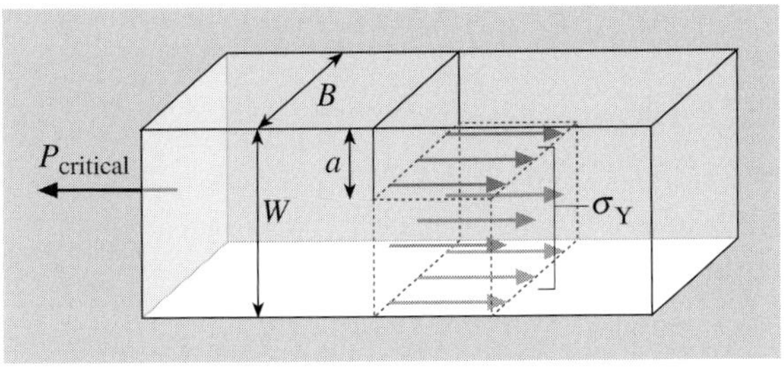

Figure 44 A lower bound analysis of a cracked bar

$$\text{force} = \text{stress} \times \text{area}$$

is the equilibrium condition, so the critical load is simply that at which the ligament becomes fully plastic at the yield stress of the material,

$$P_{critical} = \sigma_Y B(W - a)$$

because the cracked area carries no load.

This lower bound analysis turns out to be an accurate solution that is valid for small and large values of B, which is both counterintuitive and surprisingly useful for back-of-an-envelope calculations.

Imagine that B is small, so creating a thin sheet that will behave close to plane stress conditions. It is easy to imagine that the sheet fails by necking along the ligament. Figure 45 shows this effect on a section reduced by the presence of a hole.

However, the mode of failure is entirely different if B is large, so approximating to plane strain conditions at the ligament. Figure 46 shows a crack, centre top, in a thick section under tension. The surface is acid etched to differentiate regions of extensive plastic deformation. This betrays lines of plastic deformation shooting out from the crack tip at angles of about 45° to the free surface at the bottom of the picture. These are lines of intense shear along which massive displacements take place, like shearing a pack of cards. Once these bands of intense plasticity have broken to the surface the structure becomes a mechanism and plastic collapse takes place.

Figure 45 Necking of a thin plate

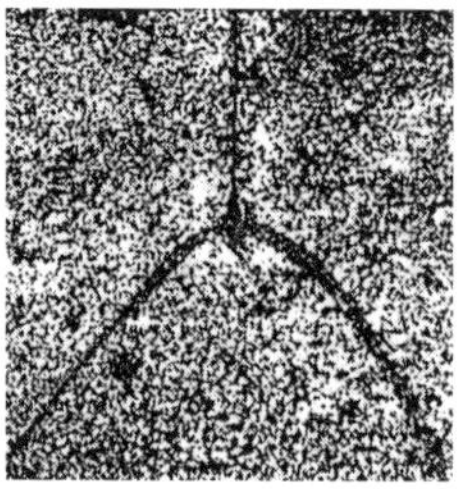

Figure 46 Bands of intense shear in a bar in tension

An upper bound analysis using this deformation mechanism produces the same numerical result as a lower bound equilibrium analysis, even though the mechanisms for plastic collapse are completely different: necking across the section in plane stress and twin shear bands in plane strain. Unfortunately this happy coincidence does not extend to other geometries and loadings.

Calculation of plastic collapse loads depends strongly on the geometry of a loaded component. Figure 47 shows a double edge cracked bar in tension under plane strain conditions. The plasticity has been contained and spread by an interaction between the two cracks.

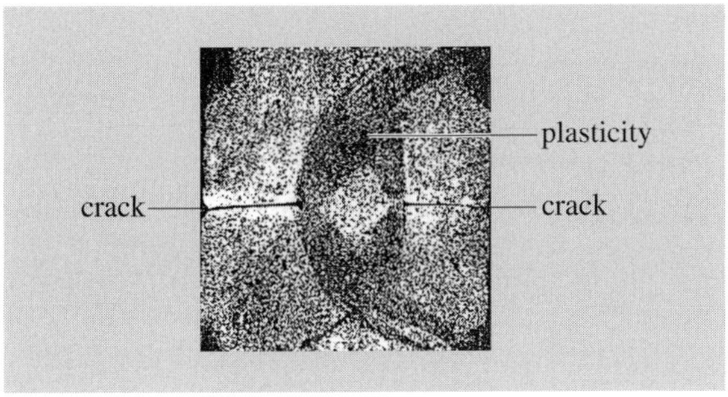

Figure 47 The presence of a second crack smears out the plasticity

In plane stress it is easy to visualize that failure will still occur by necking on a line between the crack tips. A calculation of plastic collapse loads produces a much higher value in plane strain because of the large area being plastically deformed. Handbook results for the two cases produce plastic collapse remote stresses of:

$$\sigma = \sigma_Y \left(1 - \frac{2a}{W}\right) \text{ for plane stress}$$

$$\sigma = 2.57\sigma_Y \left(1 - \frac{2a}{W}\right) \text{ for } \frac{2a}{W} \geq 0.884 \text{ for plane strain}$$

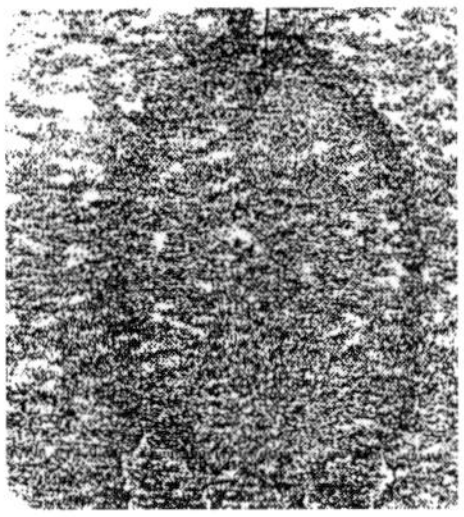

Figure 48 A band of intense shear forms a hinge in bending

In Figure 48 a short crack is shown at the top in the centre. The thick section, in plane strain, is loaded in bending. Etching the surface shows an elliptical band of intense shear that forms a 'hinge' when plastic collapse takes place.

A lower bound solution for a beam in bending requires only a stress system that is in equilibrium; it does not have to be the actual stress system, so it is sensible to choose a simple stress system for ease of analysis.

Figure 49 shows a bending stress distribution that has two blocks of stress at yield on the ligament, one in tension and one in compression.

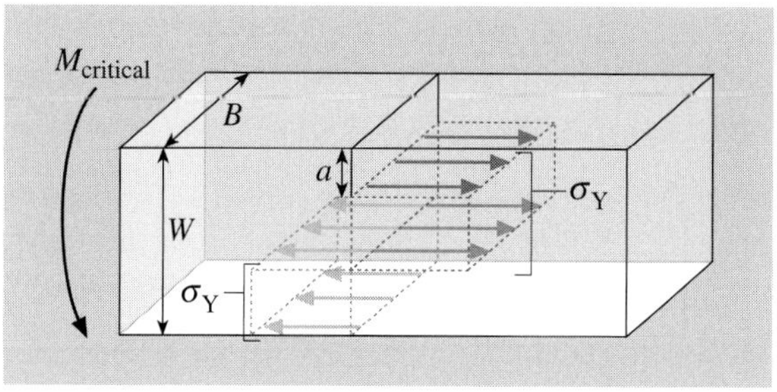

Figure 49 A lower bound analysis of a cracked beam in bending

Each block of stress is equivalent to an internal force of

$$\sigma_Y B \frac{(W-a)}{2}$$

acting at its centre. The two opposing forces form a couple a distance of

$$\frac{(W-a)}{2}$$

apart that is in equilibrium with the applied moment.

So

$$M_{critical} = \sigma_Y B \frac{(W-a)}{2} \frac{(W-a)}{2}$$

$$M_{critical} = 0.25\sigma_Y B (W-a)^2$$

where $M_{critical}$ = critical bending moment

As this is a lower bound analysis it gives either the correct result or a result that is lower than it. An upper bound analysis gives

$$M = 0.31\sigma_Y B (W-a)^2$$

So, the correct multiplier lies somewhere between 0.25 and 0.31. Modern numerical analysis techniques produce accurate solutions for very complex geometries that are available in handbooks, or can be bought. So, the problem of analysing plastic collapse loads turns into one of looking up solutions in industry-specific, validated handbooks.

In a handbook published by British Energy (Laham, 1999), the company responsible for maintaining nuclear reactors, an embedded flaw, Figure 50, is modelled as an ellipse offset from the centre line of a plate. The plate is loaded out of the plane of the paper. Here I show solutions for tension and bending loadings.

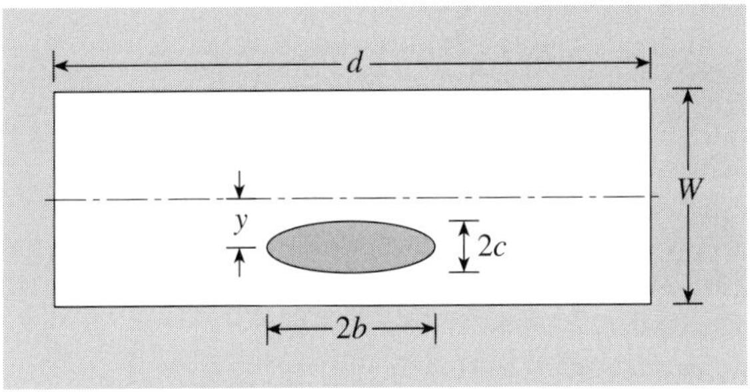

Figure 50 An embedded flaw

If the plate is loaded by an axial load, N_L, the collapse load depends on the way in which the load is applied. If the tension is applied through fixed grips then global collapse does not depend on the flaw offset, y. However, if the plate is loaded through pins so rotation can take place at the load positions, then the collapse load is a function of the flaw offset.

Solutions

Tension through grips:

$$\text{global collapse load, } N_L = \frac{\sigma_Y d}{(w+b)}\left(w^2 + b(w-2c)\right)$$

Tension through pins:

$$\text{global collapse load, } N_L = \frac{\sigma_Y wd}{(w+b)}\left(w + b\left(\left(1 - \frac{8cy}{w^2}\right)^{0.5} - \frac{2c}{w}\right)\right)$$

If the loading is by an out-of-plane bending moment, M_L, then the plastic collapse load is:

$$\text{global collapse moment, } M_L = \frac{\sigma_Y d}{4(w+b)}\left(w^3 + b\left(w^2 - 4c^2 - 8cy\right)\right)$$

If the embedded flaw is centrally located then the collapse load is obtained by putting y equal to zero:

$$\text{global collapse moment, } M_L = \frac{\sigma_Y dw^2}{4(w+b)}\left(w + b\left(1 - \frac{4c^2}{w^2}\right)\right)$$

In addition to loading constraints, handbooks provide solutions for local and global collapse. In Figure 51 a defect is shown on the inside wall of a pipe under pressure, in order to illustrate the sort of issues that need to be dealt with when using handbook solutions.

Collapse mode	Pressure on crack faces	Handbook solutions for plastic collapse load
Global	✗	$\sigma_Y\left[\dfrac{c}{r_1\left(1+\dfrac{1.61b^2}{r_1c}\right)^{\frac{1}{2}}}+\left(\dfrac{r_1}{r_1+c}\right)\ln\left(\dfrac{r_2}{r_1+c}\right)\right]$
Global	✓	$\sigma_Y\left[\dfrac{c}{r_1\left(1+\dfrac{1.61b^2}{r_1c}\right)^{\frac{1}{2}}}+\left(\dfrac{r_1}{r_1+c}\right)\ln\left(\dfrac{r_2}{r_1+c}\right)\right]$
Local	✗	$\dfrac{\sigma_Y}{2(s+b)}\left[s\ln\left(\dfrac{r_2}{r_1}\right)+2b\ln\left(\dfrac{r_2}{r_1+c}\right)\right]$ where $s=\dfrac{bc\left(1-\dfrac{c}{t}\right)}{r_1\left(1+\dfrac{1.61b^2}{r_1c}\right)^{\frac{1}{2}}\left[\ln\left(\dfrac{r_2}{r_1}\right)-\ln\left(\dfrac{r_2}{r_1+c}\right)\right]-c}$
Local	✓	$\dfrac{\sigma_Y}{2(s+b)}\left[s\ln\left(\dfrac{r_2}{r_1}\right)+b\left(\dfrac{r_1}{r_1+c}\right)\ln\left(\dfrac{r_2}{r_1+c}\right)\right]$ where $s=\dfrac{bc\left(1-\dfrac{c}{t}\right)}{r_1\left(1+\dfrac{1.61b^2}{r_1c}\right)^{\frac{1}{2}}\left[\ln\left(\dfrac{r_2}{r_1}\right)-\left(\dfrac{r_1}{r_1+c}\right)\ln\left(\dfrac{r_2}{r_1+c}\right)\right]-c}$

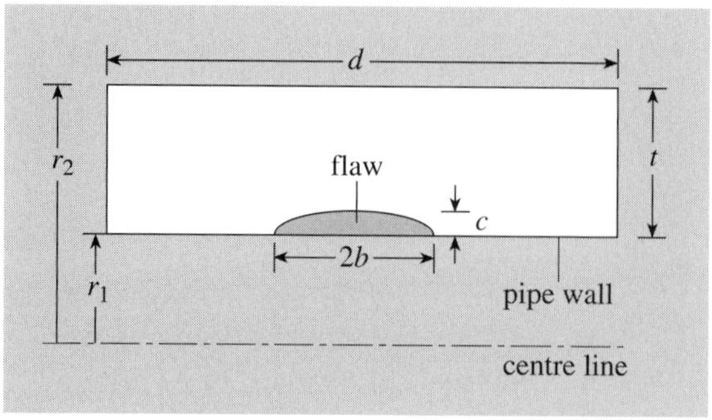

Figure 51 Wall defect in a pipe under pressure

The first problem is to decide whether plastic collapse will be a global or local event. If the collapse is local, a bubble will be blown on the outside of the pipe at the flaw site. Global collapse will be seen as a circumferential bulge, also at the flaw site. A real failure might well lie somewhere between these two extremes.

The second problem is to decide whether the internal pressure is forcing the crack faces apart or just stressing the pipe wall, for example if the crack is blocked by paint or rust. A real failure might well lie somewhere between these two extremes. Although these equations contain a number of terms they are simple to calculate using a spreadsheet or proprietary software. The usual procedure of performing a failure assessment against the most conservative assumption is straightforward. However, if a lower bound assessment fails, a 3D finite-element analysis might be needed to analyse the intermediate case between local and global plastic collapse.

Elastic plastic fracture mechanics (EPFM)

Crack initiation resulting in fast, brittle fracture on the one hand, and plastic collapse without growth of a pre-existing crack on the other, bound all the possible fracture events that can take place in a metal. In between these two extremes lie more complex possibilities, here arranged in order of decreasing danger:

- Unstable ductile fracture leaving behind a torn fracture surface.
- Slow, stable, ductile tearing followed by brittle fracture.
- Slow, stable, ductile tearing followed by ductile fracture.
- Slow, stable, ductile tearing followed by plastic collapse.

The complexity of these processes poses significant theoretical and practical challenges. The nuclear industry took up these challenges, starting in the US and later in the UK and France, reflecting the maturity of their respective nuclear power generation programs.

This industry had the problem of developing a community of practice and of research whilst building and commissioning new reactors, at the same time as coping with ageing and deteriorating reactors.

The industry's safety-conscious culture developed conservative design and analytical processes for itself and its suppliers under a regime of government regulation. The CEGB in the UK, for example, needed to develop formal assessment methods and to communicate safe custom and practice to engineers in a wide range of supply industries. It is this that drove the development of their two-parameter, R6 failure assessment diagram (FAD). This powerful method of representing failure uses the two extremes of brittle fracture and plastic collapse to create a failure map.

Why R6? The original report came out of the central R&D offices of the CEGB because their laboratories at Berkeley and Leatherhead were unwilling to concede reporting from any other lab. After twenty years central R&D had only produced five reports and this was the sixth. Its production coincided with a senior manager returning from Germany with a special duty-free bag carrying an R6 logo containing a pack of R6 cigarettes into which he had put two packs of R3 contraceptives.

The first version of R6 *Assessment of the integrity of structures containing defects* was published in August 1976 and revised in April 1977 to support tender documents associated with reactor design. Procedures for dealing with secondary and residual stresses were introduced in 1980.

A major revision, R6 Revision 3, introduced a new mapping in 1986 to take into account the work-hardening characteristics of metals, particularly of strongly strain-hardening austenitic steels. R6 Revision 4 was introduced in 2001 to simplify the basic procedures, and to complicate the treatment of secondary stresses.

Despite its uninspiring name this document has lasted and has influenced the development of British standards, most notably BS 7910 which contains much of Revision 3. When the CEGB was dismantled in 1989 ownership of the code moved to Nuclear Electric, and subsequently to British Energy Generation who maintain and develop it in conjunction with Serco Assurance (formerly AEA Technology).

The emphasis in R6 has always been on failure avoidance within the type of safety case procedures outlined earlier. However, the representation is so powerful that using it for failure prediction is almost inevitable.

The next section introduces the ideas behind R6, and tracks its development to the all-important inclusion of work-hardening properties.

Mapping failure

Recognizing that plastic collapse and brittle failure are orthogonal properties drove engineers at the CEGB to develop the idea of displaying failure as a map of this space.

For all failure modes there is a mechanics driving parameter and a materials resistance parameter. For example, stress is a mechanics parameter and it increases until it reaches the yield strength (a materials parameter) when plastic deformation occurs. So, by making the axes a ratio of the relevant mechanics parameter to the relevant materials parameter, failure is predicted when the ratio equals 1. In effect, the axes are a dimensionless measurement of nearness of approach to failure: the closer the measure is to 1, the nearer it is to failure, Figure 52.

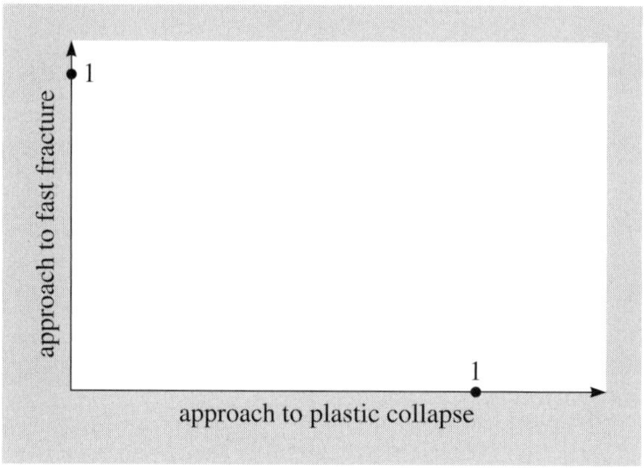

Figure 52 Orthogonal axes for distinct failure modes

The CEGB used

$$K_r = \frac{K}{K_c}$$

for approach to fast fracture, and

$$S_r = \frac{\sigma}{\sigma_1}$$

for approach to plastic collapse but, in principle, any measure of an appropriate materials property can be used.

K is the stress intensity factor due to the applied load. K_c is a measure of material fracture toughness, the lowest value of which is K_{IC} and so this is the most conservative measure.

The stress due to the applied load is σ. The stress at which the structure becomes fully plastic is σ_1, which is calculated from the flow stress of the material and the geometry under load. The flow stress, which is halfway between yield stress and ultimate stress of the metal, is a practical, compromise value in general engineering use. This is an arbitrary choice, as there is still a residual load-carrying capacity when the flow stress is reached. The yield stress, which would have increased conservatism, could just as easily have been chosen for the development of the method.

The problem is what to do about the space in between brittle failure and plastic collapse, which is the domain of elastic plastic fracture mechanics (EPFM). Here there are a number of possible types of failure that depend critically on the extent of plasticity in the section of a component or test piece and the hardening characteristics of a material. The extremes are relatively uncomplicated but the possibilities of different types of failure in the intervening space are complex, from both a theoretical and a practical viewpoint. So, what to do?

The Dugdale 'strip yield' model was published in 1960. This theory models a crack in a thin plate with a planar strip of yielded material ahead of the crack, Figure 53.

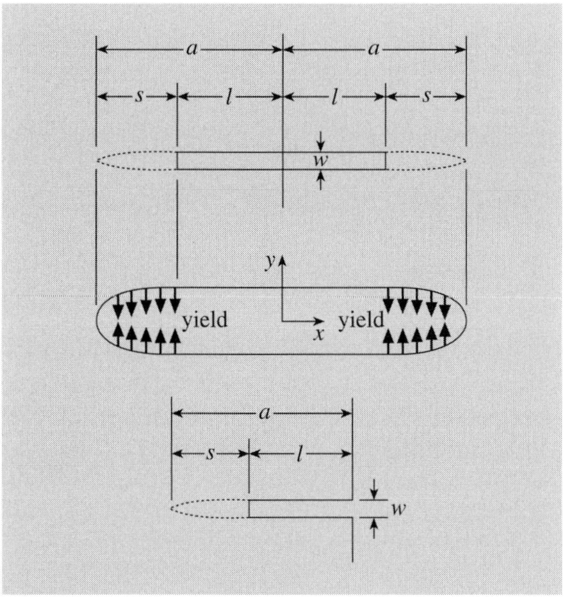

Figure 53 Modelling the scale of the plastic zone

Dugdale's model imagines a crack that is longer than the real crack by the size of its plastic zone. This virtual crack is loaded by the remote opening stress and a stress of σ_Y that pinches the extra length closed. These two loadings produce elastic singularities that can be taken to cancel each other out. Therefore a 'plastic' analysis is created from two elastic analyses to model the scale of the plasticity, even though no attempt has been made to model its shape accurately. The analysis produced by this model joins the brittle fracture condition to the plastic collapse condition.

At the CEGB they used a variation of this model to produce the equation

$$K_r = S_r \left[\frac{8}{\pi^2} \ln \sec \left(\frac{\pi}{2} S_r \right) \right]^{-1/2}$$

which joins the K_r and S_r axes to create a failure envelope, Figure 54.

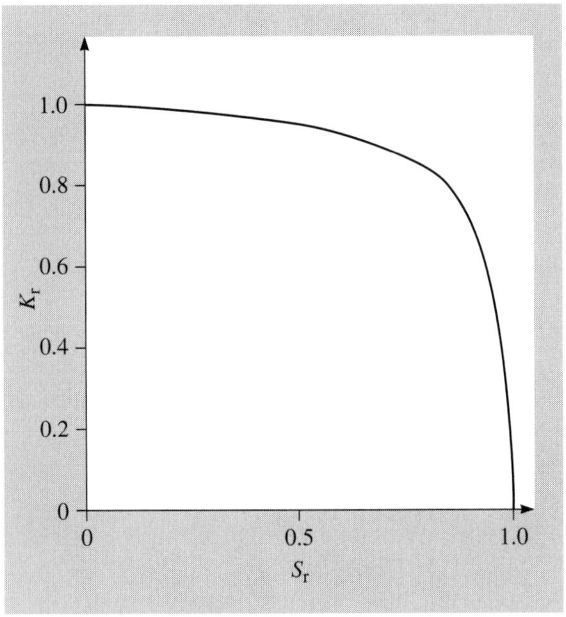

Figure 54 The R6 Revision 2 FAD: safe operation for points within the envelope

K_r labels the LEFM axis:

$$K_r = \frac{K}{K_{IC}}$$

S_r labels the plastic collapse axis:

$$S_r = \frac{\sigma}{\sigma_1}$$

To use an R6 diagram requires a stress analysis, knowledge of the length of a crack, three materials properties and access to reference handbooks.

The handbooks provide a plastic collapse relationship and a K calibration relationship for the crack geometry being investigated.

The stress analysis provides the remote stress loading on the crack that, together with its length, enables K to be calculated from the K calibration relationship.

The first material property required is the material's fracture toughness, K_{IC}.

Hence

$$K_r = \frac{K}{K_{IC}}$$

The plastic collapse handbook provides the formula for calculating the stress that causes full plasticity at the ligament from the geometry and loading of the body. The yield stress and ultimate tensile stress are needed to calculate the flow stress.

Hence

$$S_r = \frac{\sigma}{\sigma_1}$$

The two values produced create the coordinate point (K_r, S_r) shown in Figure 55.

The nearness of approach of the coordinate to the failure envelope is therefore a measure of the reserve against load, or safety factor.

Because K_r and S_r both change with the stress, which changes with load, any increase in load drives the coordinate point towards the failure envelope boundary along a line from the origin.

Any number of interesting points could be plotted along this line. For a pressure vessel this might be plotting a normal design working load, an occasional high design load caused by, say, starting from cold, and the proof load used before commissioning the vessel. These would produce a vivid representation of the span of loading cases along a line on one diagram.

Increasing the load on the structure represented by Figure 55 drives a point in the centre of the map towards an eventual failure into the EPFM region, away from brittle failure or plastic collapse. Figure 56 shows that a coordinate point close to the K_r axis will be driven towards brittle failure at the boundary by an increase in load, and hence stress. A coordinate point close to the S_r axis will be driven towards plastic collapse at the boundary as the loading is progressively increased.

The coordinates of a point on the diagram represent values calculated from important engineering parameters such as load, crack length, and so on. Imagine that the point is being driven around the map by changes in these variables.

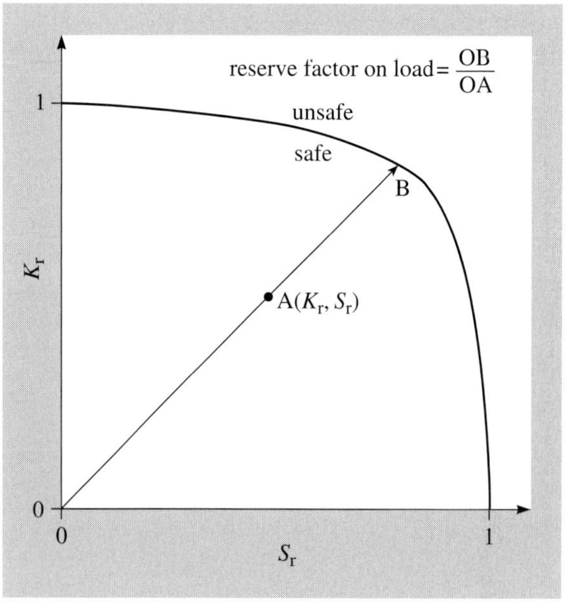

Figure 55 Plotting a coordinate on the FAD

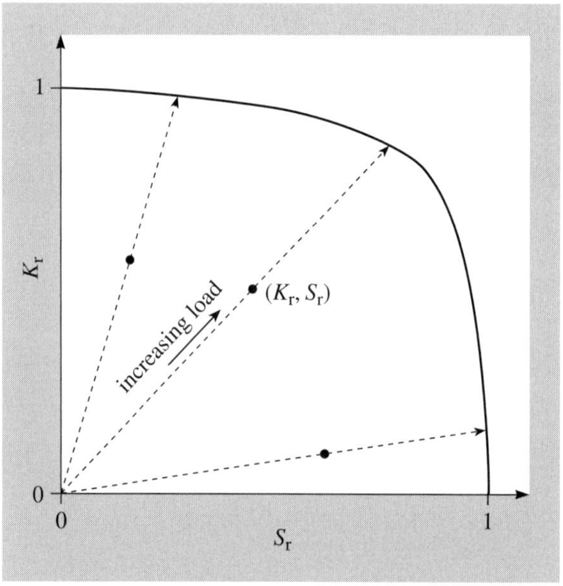

Figure 56 Approaching the failure envelope in different zones

It is easy to imagine a change in any important engineering variable due to, say, an improvement in material specification, or the relaxation of some conservative assumption. Figure 57 shows the paths traced by the coordinate point when fracture toughness and yield stress are increased. Note that from the new positions of the coordinate points a subsequent increase in stress will drive the point towards different regions on the map.

For example, if the material is tougher the point is driven towards failure by plastic collapse, whereas an increase in yield suggests an intersection at the boundary in the brittle region.

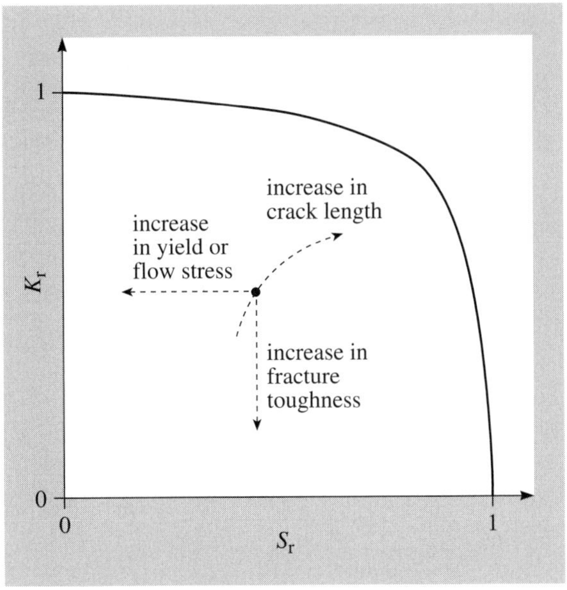

Figure 57 Changing the driving force

It would be valid to use the FAD in conjunction with information about crack detection sensitivity; for example, to estimate the significance of being able to guarantee finding, say, a 5 mm or 3 mm embedded crack using X-ray or ultrasonic non-destructive examination techniques. Figure 57 also shows that an increase in crack length does not produce a linear response on the map, because *K* changes like the square root of crack length, not linearly as with the other parameters considered.

An issue in the use of FADs over the years has been the treatment of residual, or secondary, stresses. Here I shall illustrate the problem briefly by considering local stresses introduced by welding.

Welding is a form of local casting by melting and filling gaps between neighbouring, solid metals. When the weld metal solidifies it contracts, so setting up local stresses around the weld. These stresses must be in local equilibrium so some will be tensile and some compressive.

Cast weld metal and the neighbouring heat-affected zone (HAZ) may have poorer material properties than the surrounding, forged, parent plate. In addition, the welding process is very good at introducing flaws. When there are flaws in a material with reduced toughness, and there are also residual stresses present that combine with loading stresses to open, say, an embedded crack, the results can be impressive: look back to Figure 23.

Conservative assumptions in creating a safety case would demand that the flaw and the stresses are all oriented to produce a worst case.

There are two major problems in dealing with residual stresses: the first problem is being able to determine their value, and the second is what happens to them in the presence of local plasticity.

Residual stresses in the body of a metal are difficult and expensive to measure, and hard to eliminate. Best practice demands pre- and post-weld heat treatments to reduce the propensity for introducing stresses and ameliorating the value of those introduced. In thick sections it is virtually impossible to eliminate welding residual stresses completely. The ballpark figures for residual stresses in a pre- and post-weld heat-treated vessel are about 20% of yield, and for a vessel without treatment about 80% of yield.

Conservative failure assessment procedures often require the starting assumption of residual stresses at yield and that measurements of residual stresses above yield have been made in highly constrained geometries. These are very high values of residual stress when compared with the figure for a load-induced design stress of two-thirds yield for a highly loaded structure.

Local plasticity during shakedown and proof loading is relied on to 'wash-out' self-equilibrating residual stresses; but how much residual stress has been washed out and how much remains at some lower value of load?

Think about it this way. Imagine, as a starting point, that you have a material that exhibits hardly any plasticity at all. Therefore a residual stress and a loading stress simply add together to drive a flaw to brittle fracture. You have a 'control knob' that changes the ductility of the material by decreasing the flow stress. Increase the material's ductility a little and some shakedown occurs as the load increases. This reduces the self-equilibrating residual stresses by a modest amount, but perhaps not enough to prevent brittle fracture. Now give the material a decent amount of ductility. Most of the residual stresses are washed out as the load increases, and the flaw might tear in a ductile fashion. Turning the knob to maximum washes out the entire residual stress component as the load is increased, so leaving the loading stress to drive the structure to plastic collapse.

The easiest way to represent the effect of residual stresses on the FAD is shown in Figure 58. The residual stress has been added to the loading stress for the calculation of K, but is assumed to make no contribution to plastic collapse. The effect of this is to move the load line up the diagram. This is easiest to imagine at zero external load; with no loading stresses there remains a residual stress to provide a K driving force so the load line no longer goes through the origin. This applies to all points on the load line, which simply moves up the diagram. Unfortunately this approach is not necessarily conservative in the EPFM regime because it does not take account of the residual stresses that remain to drive failure. Various schemas to take residual stresses into account have been introduced into R6 over the years. As the values of residual stresses are not well known in the first place this is a particularly frustrating business. Therefore the simple technique illustrated here is a good enough starting point. It is in the area of secondary stresses that Revision 4 of the code has become more complicated, in particular by introducing a new class of medium-range secondary stresses.

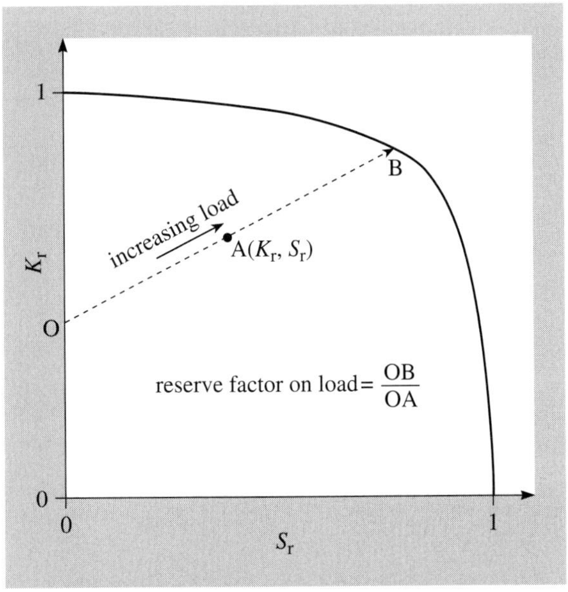

Figure 58 Residual stresses displace the load-line upwards

Although this FAD is a superb aid to communication its theoretical basis is not a close fit for fractures of tough materials in smaller sections. So, as ever, practical engineering imperatives demand that such procedures are subject to trial-by-data in order to validate the technique.

When experiment met theory

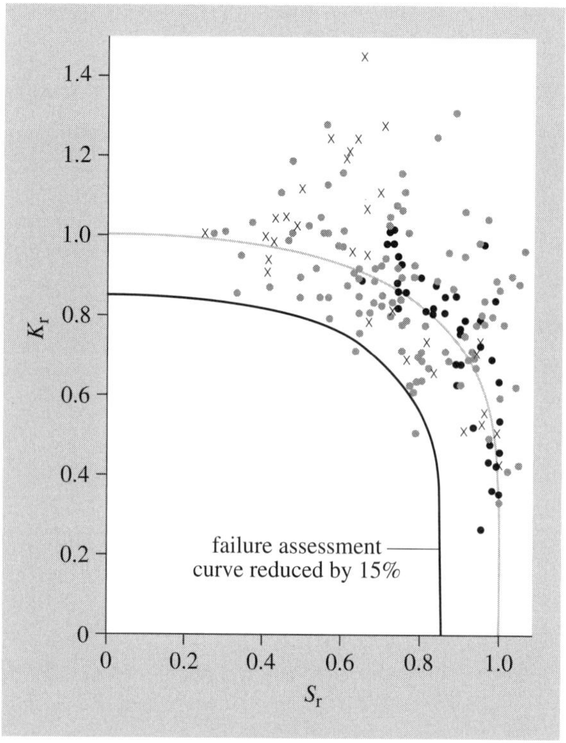

Figure 59 Theory meets practice

Graham Chell published the diagram shown in Figure 59 in 1979 to validate the FAD. It contains a wealth of experimental data, collected on aluminium and titanium alloys, and a range of different steels in various states. Size effects, crack lengths and loading conditions also contributed to the rich variety of engineering conditions portrayed by these data.

Figure 59 also shows a lower bound curve constructed to put these data outside of the failure envelope created by the strip yield model.

There are other, interesting observations that can be made from the distribution of data on the diagram. In particular, notice that most failures occur in the EPFM region where the scatter is the greatest, that there are few data in the brittle region and that the data in the plastic collapse region shows little scatter.

Some at the CEGB argued that discrepancies between theory and practice were due to materials variability.

> *The experimental data scatter about the failure assessment curve. The principal reason for this is materials variability, the assessment points*

being determined using average values of toughness and tensile properties.

This is somewhat disingenuous given the simplicity of a materials representation that does not model work hardening. After all it is in the EPFM region that the detail of work hardening has most effect, and it is there that the experimental scatter is at its greatest. This omission was the subject of a heated discussion at a 1979 Royal Society meeting on fracture mechanics. However, the Revision 2 diagram was a good starting point for introducing a new technique to the engineering community and it was inevitable that such a powerful, pictorial representation of fracture would develop over the years.

To take the subject further needed a new theoretical formulation. The foundations of a theory for EPFM were in place at the time, but the incorporation of theory into routine practice is a necessarily slow process driven by need, cost and confidence.

J is for Jim

The story of elastic plastic fracture mechanics (EPFM) parallels that of LEFM in the sense that there is an energy representation and a stress representation of the problem.

Figure 60 Jim Rice

In 1968, Jim Rice (Figure 60) produced an expression based on the strain energy distribution and the displacement along a path around the crack tip that he called the *J*-integral. *J* can be thought of as the rate of change of work input to a material as a function of crack growth that is measurable at the loading pins of a cracked specimen.

He had already identified *J* as a parameter that characterized the strength of the concentrated stress–strain field at a notch or crack tip. However, he downplayed the energy interpretation until the stress formulation was established.

> *The reason for putting the energy interpretation last is that I sensed that it would cause ceaseless confusion, as turned out to be the case, since the energy interpretation applies for a growing crack in a nonlinear elastic material but not in an actual, incremental elastic–plastic material.*

The stress-based representation of a crack in the presence of significant plasticity was established when Hutchinson, and Rice and Rosengren, related the *J*-integral to the crack-tip stress fields in non-linear, elastic materials in the same edition of the *Journal of the Mechanics and Physics of Solids*.

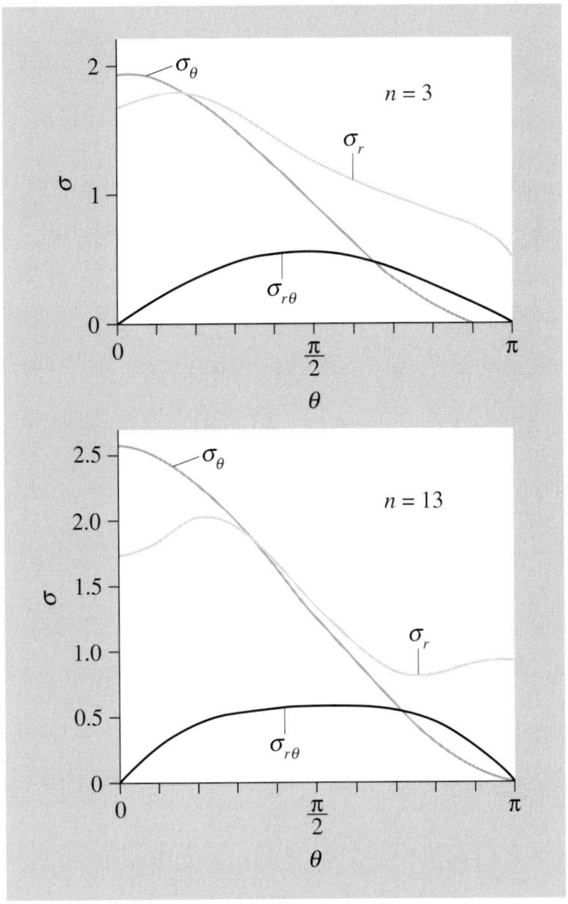

Figure 61 Another stress field

As with the K field there is a singularity, now known as the Hutchinson, Rice and Rosengren (HRR) singularity, which creates the same sort of anomaly as its LEFM equivalent because the stress is predicted to reach infinity at the crack tip. The expression that produces the pattern shown in Figure 61 is

$$\sigma_{ij} = \sigma_Y \left[\frac{EJ}{\alpha \sigma_Y^2 I_n r} \right]^{\frac{1}{n+1}} f_{ij}(\theta, n)$$

where I_n depends on the work-hardening coefficient, n. They are Hutchinson's original numerical calculations of the stress field variation for two different values of n. The field is valid only for materials that can be modelled by a power-law relationship. Note that for n = 1, Figure 12, the field returns to the $r^{-1/2}$ singularity of LEFM theory.

So, for a linear elastic material,

$$J = \frac{K^2}{E}$$

As with K there are complicated geometrical effects to be considered. However, the most important additional complication is the inclusion of work hardening in the model using a non-linear elastic model of plasticity. In this model the load–deflection curve traces a curved path on loading, but retraces the same path on unloading so there is no unrecoverable work, as there is in the deformation of a plastic metal.

J can also be thought of as a 'magnification factor' for a stress pattern in the same way as K. The idea of crack growth that comes out of J is akin to a series of stills from a celluloid filmstrip. Each frame is a frozen picture of the mechanical state of the material close to the crack tip. What happens between the frames is left to the imagination.

As well as ignoring the irrecoverable plastic deformation, this theory does not model crack blunting or the large strain effects close to the crack tip. It is these effects that destroy the HRR singularity, as does plasticity for the LEFM singularity.

Limits to LEFM and EPFM models

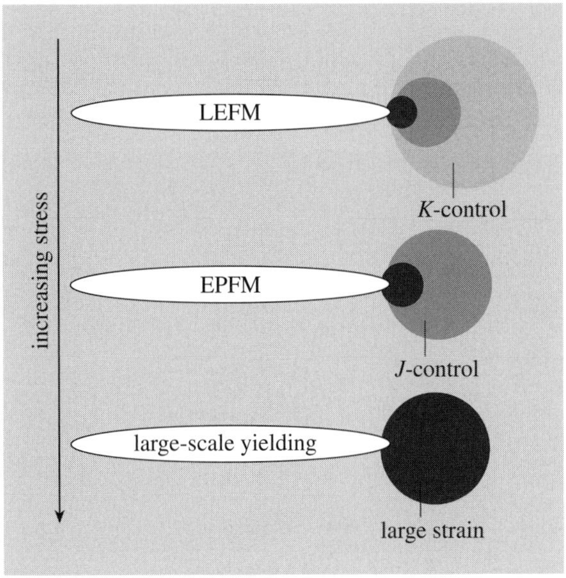

Figure 62 Little and large

Under conditions of small-scale yielding there are large strains within an HRR stress field in the plasticity ahead of a crack tip, as shown in the top diagram of Figure 62. These are unimportant whilst the plasticity is small because the surrounding *K* field can be used to characterize a brittle-fracture initiation event. This is known as *K*-control, LEFM, or small-scale yielding.

When the plastic zone becomes too large the *K* field is no longer a valid crack-tip characterizing parameter. In the middle diagram of Figure 62, the HRR field takes the place of the *K* field as the characterizing parameter. It encloses a large strain region that destroys the predicted HRR singularity. This is known as *J*-control or EPFM.

When the large strain region around the blunted crack tip dominates, the *J*-control field is no longer a crack-tip characterizing parameter, as shown in the bottom diagram of Figure 62.

Figure 63 shows numerical results of stress against distance from the crack tip for a small, three-point bend specimen and a massive, centre-cracked panel in tension. Solutions for *K* and HRR fields are shown in the same space for a *K* of 35 'fracture mechanics units' loading both specimens.

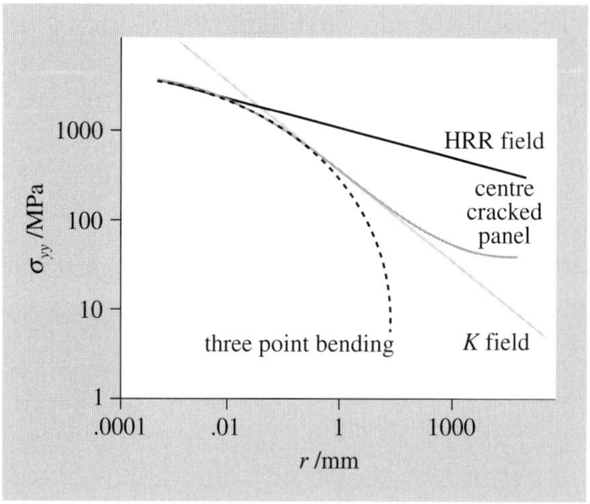

Figure 63 Comparing numerical analysis with theory

The HRR field matches both numerical solutions close to the crack tip. Further out from the crack tip the very large centre-cracked panel's stress field is well described by the *K* field for a long distance, but the much smaller bend specimen kisses the *K* field for a short distance only.

Since the 1980s numerical methods have become dominant in engineering practice. These methods, which can cope with the difficulties of complex geometries, loadings and plasticity, are used to determine the validity and range of application of those theories developed between the 1950s and the 1970s.

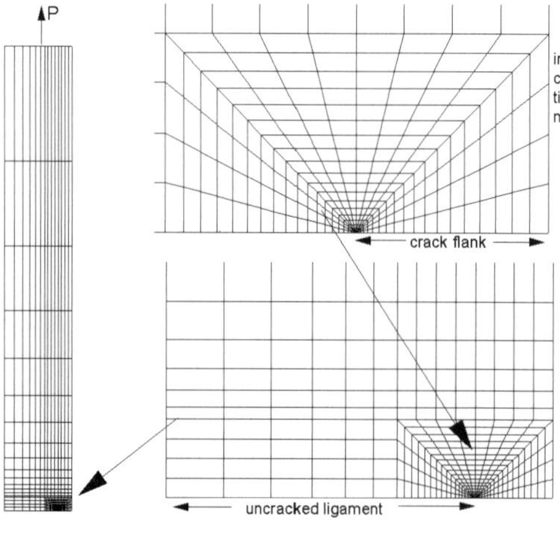

Figure 64 A finite-element mesh

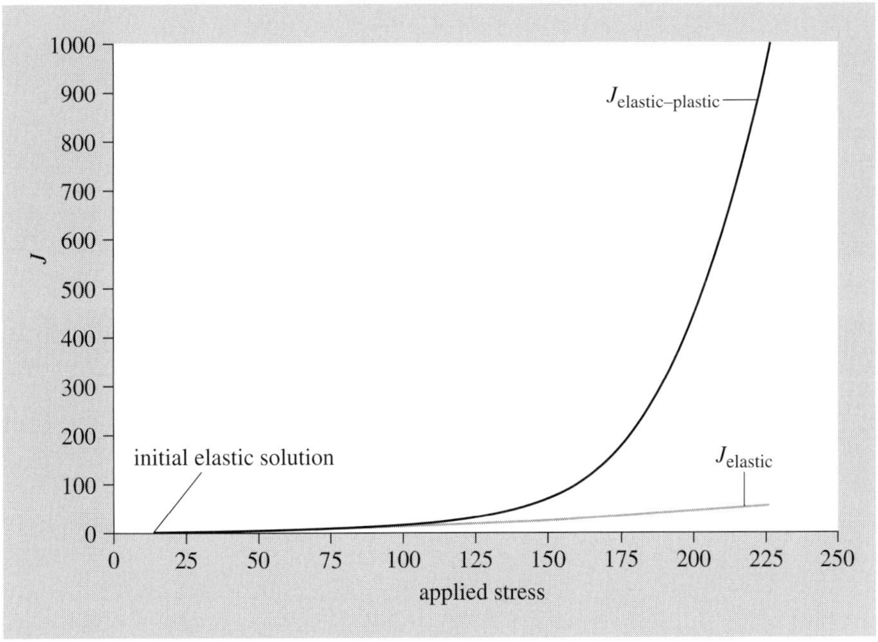

Figure 65 Results from the finite-element analysis

A finite-element analysis divides the space being analysed into elements connected at nodes. Figure 64 illustrates a finite-element mesh created to model a crack. Figure 65 portrays the results, showing how an analysis that incorporates plasticity deviates from a linear elastic analysis.

$$\sqrt{\frac{J_{\text{elastic}}}{J_{\text{elastic-plastic}}}}$$

can be used in conjunction with a measure of material toughness expressed in terms of J to represent a failure boundary. By this means a direct link to the R6 method can be made.

The reason for the square root is the use in the UK of K for historical reasons. Remember that

$$K_{\text{r}} = \frac{K}{K_{\text{material}}}$$

where K is the LEFM stress intensity factor. K can be easily replaced by

$$\sqrt{J_{\text{elastic}}}$$

or more formally

$$K = \sqrt{\frac{EJ_{\text{elastic}}}{\left(1-\nu^2\right)}}$$

Perhaps it would have been clearer if the *K* terminology had been left in the past, as a special case of *J*, but a generation's worth of terminology is not thrown away lightly.

The *J* procedures, colloquially known as GE Scheme, were originally formulated at the General Electric Company in the USA, and crystallized in a document prepared for the Electric Power Research Institute (EPRI) in 1981 (Kumar et al.).

From the engineer's viewpoint the really good news is that it is not necessary to analyse EPFM problems from scratch, because *J* is contained implicitly in a revised R6 FAD.

Revising the failure assessment diagram

It fell to the 1986 Revision 3 of R6 to introduce the effect of hardening, a major change to the code. Figure 66 shows the most important FAD curve from both the 1986 and 2001 revisions. It is comforting to see the curves so close together because it suggests that over half a decade there were no nasty surprises to disturb a mature procedure. However, the curve is a very different shape from the Revision 2 curve and the terminology is different. So there is something new to learn. The good news is that the procedure is used in the same, simple way and that it relieves the user from having to perform or buy a J analysis.

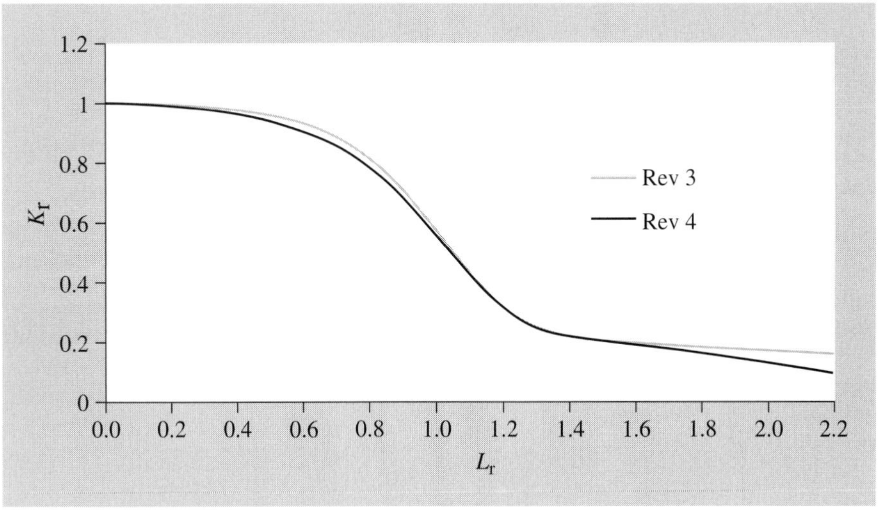

Figure 66 The lower bound FADs of 1986 Rev 3 and 2001 Rev 4

The curve shown is the lower bound to a number of materials and structure-specific curves created by conducting J analyses. These materials-specific curves were created from the particular stress–strain curves of the material being analysed, together with some pessimistic assumptions to remove geometry-dependent effects based on a testing program. Three-point bending, compact tension specimens and centre- and double edge-cracked panels were tested at three different crack depths. Specimen width effects were also assessed by testing a pressure vessel steel in bending at the extremes of 10 mm and 230 mm wide.

This lower bound curve provides the confidence of a theoretical basis much improved over its predecessor, together with support from an experimental testing program. To produce the lower bound curve shown in Figure 66, that covers a range of metals and is geometrically insensitive, required many years of effort and development. Implicit in a lower bound curve is a J analysis. The potential usefulness of the tool as a failure predictor is therefore considerably enhanced.

The horizontal axis still measures nearness of approach to plastic collapse, but in terms of load rather than stress.

$$L_r = \frac{P}{P_Y}$$

where P is the primary load (no secondary stresses). P_Y is the load that causes plastic yield of the cracked section, but based on the yield stress of the material, not on the flow stress as in Revision 2. In Revisions 3 and 4 the flow stress is put into the curve, not into the assessment coordinates. Notice that the lower bound curve shown in Figure 66 does not have a definite end point; this is provided by a cut-off calculated as the ratio of the flow stress to the yield stress for the material being considered. Figure 67 shows the effect of putting in a maximum value of

$$L_r = \frac{\sigma_{\text{flow}}}{\sigma_{\text{yield}}}$$

for two materials with very different work-hardening characteristics.

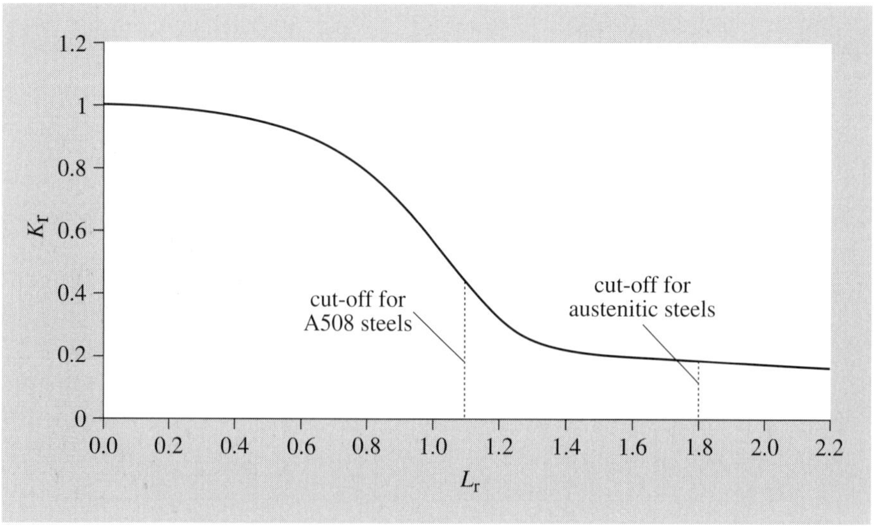

Figure 67 Modelling flow stress on the lower bound curve

The development of EPFM ideas introduces the possibility of altering the value of K_{material}. The plane strain fracture toughness for brittle fracture, K_{IC}, is the only materials property available to describe brittle, fast fracture. In the EPFM regime a crack initiation event needs to be defined in order to represent the start of failure. As with equating a K crack driving parameter to a K materials parameter, matching a J crack driving parameter to a J materials parameter is the usual procedure. International standards define the way in which a J test can be performed in order to produce a valid materials property.

In the EPFM regime, under steadily increasing load a stationary crack tip blunts under the effect of large strains until the crack grows at the tip by void coalescence. The exact point of initiation is near impossible to measure in a routine engineering test. So, a practical engineering approximation of 0.2 mm of crack growth that combines crack blunting and initiation is used to produce a value of J which is labelled $J_{0.2}$, as a materials property. There is nothing special about 0.2 mm; 0.15 mm is also used and, if electrical measurements of crack extension during a test are made, more accurate smaller values might be specified. Staying with historical terminology has its price, as this now has to be expressed in terms of K as

$$K_{0.2} = \sqrt{\frac{E}{\left(1-\nu^2\right)} J_{0.2}}$$

so extending the K terminology beyond that of the LEFM regime.

$K_{0.2}$ will be significantly higher than K_{IC} so

$$K_r = \frac{K}{K_{0.2}}$$

will be much lower than

$$K_r = \frac{K}{K_{IC}}$$

which drives the coordinate point in an R6 diagram further away from the boundary. As is customary in the nuclear industry the more conservative measure is used first, and the less conservative used only if the case fails at the first hurdle.

Plotting an assessment point on the FAD and illustrating the effect of different assumptions, changes in specification, or data from measurement is similar to the way it was discussed and illustrated in Figure 57 and Figure 58 for Revision 2. The only difference is that changing the yield stress and UTS moves both the coordinate point and the cut-off on the curve, Figure 68.

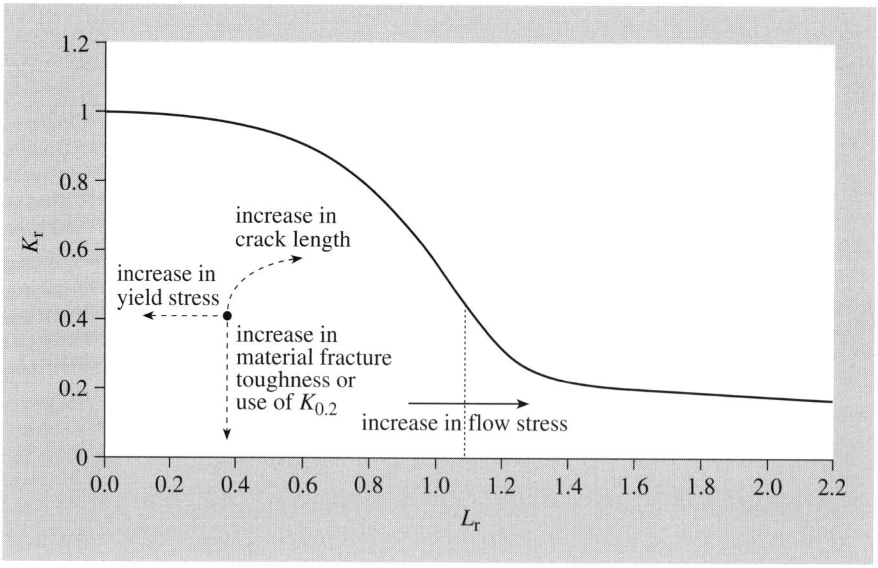

Figure 68 The effect of changing parameters on a coordinate and on the curve

Finale

The theme of this book has been that the development of engineering procedures, techniques, and tools is driven by industrial problems of significant economic importance. Industrial sectors can be as different from one another technically as they are economically. The aircraft and nuclear industries, by way of extreme examples, approach structural integrity problems very differently. Therefore, trained engineers rarely move between these two worlds of manufacture.

However, the same physical and technical principles form the foundation on which the formal procedures used by these different industries are built. Specialist materials engineers easily recognize the unifying features and mechanisms of different alloys, phases and particles, and different slip systems and types of dislocation at a microstructural level. Mechanical engineers ignore this detail to use numbers, such as values of yield stress that depend on these processes, in stress calculations to produce and to validate designs.

So, why can't engineers trained on these unifying theories move easily between different industries? It is because engineering practice is not simply the application of predictive technical theories, or, worse still, the 'appliance of science'. The procedures of engineering practice incorporate, and are limited by, their industrial contexts. These include the use of experimental data and theoretical approximations to condition a procedure to be valid for the set of products an industry makes. Properly validated procedures are applied by 'persons of competence' in the industry; Figure 69 perhaps?

The R6 procedures are a good example of how techniques grow to be more complex than their physical basis, and how quite limited physical models can form the basis of a powerful and long-lasting industrial technique.

In the heavy industries conservatism is the watchword. The results of failure in chemical, pipeline, storage and power-generation industries can be devastating: Flixborough, Three Mile Island, Bhopal and Chernobyl.

In these industries engineers implicitly take advantage of the safety margin provided by physical processes not modelled in their procedures. From a conservative starting point engineers need to be dragged, kicking and screaming, towards the use of more complex models because they are less well validated, more expensive to perform and, most importantly, erode that reassuring safety margin. I draw comfort from the slow development of industrial schemas in dangerous industries. However, it is

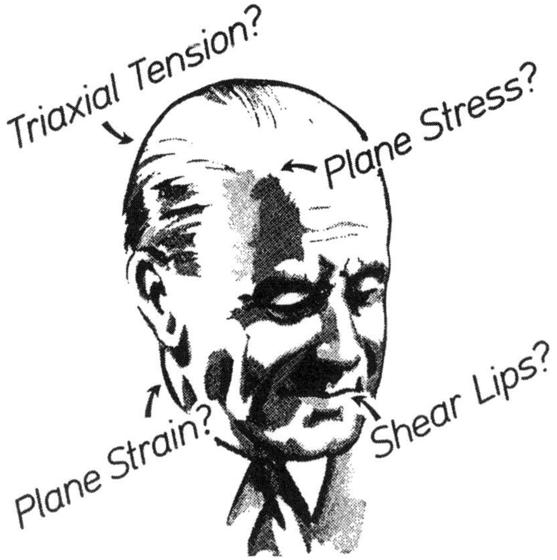

Figure 69 Headache? Take R6

a mistake to believe that procedures are logically planned and introduced, as this quote from Ian Milne illustrates.

> *We did not realise the significance of what we were doing for others. For example, we needed to produce collapse solutions since we learned that few practitioners would know how to do this. We simply made a guess at something which we thought would be pessimistic, therefore acceptable ... And having done it, it became a standard used in papers by all and sundry, quoting us as the fount of wisdom ... Never forget that if all else fails, rather than derive their own people will look for the easy way out, and pass the blame. These then become formal validated solutions written into the annals, even though they aren't.*

In the aircraft industries there is much less of a safety margin than in the heavy industries. Conservative theories that do not predict eventual failure cannot be used because aircraft need to be light enough to fly.

Major fatigue cracking of primary structure in a large fleet of commercial aircraft halfway through an aircraft's design life leads to a culture of continuous monitoring, and the incorporation of the monitoring of crack growth into design codes and procedures. It also leads to the testing of complete structures and testing during service in order to support aircraft inspectors around the world.

The Joint Airworthiness Authorities (JAA) set civil aircraft safety factors for Europe and the Ministry of Defence sets military aircraft safety factors for the UK. Military aircraft are designed with a wider flight envelope than civil aircraft. Therefore their design calculations start with a basic safety factor of 3.33 to account for scatter in experimental data. Using this factor would require the design to be supported by a full-scale, or large-component, fatigue testing, and the structure to be monitored. Without monitoring, by some sort of 'fatigue meter', the safety factor specified is 5. These factors are doubled when there is no test support for the design calculations.

The success of custom and practice in the aircraft industries is illustrated by the rarity of structural failure in the field. The Comet disaster is still the classic example of a failure in the design-test-build-monitor procedures of the industry. The Comet crashes set the UK aircraft industry back a decade in engineering and commercial competences.

This is not to say that structural failures do not occur, as the Aloha accident illustrates. The structural failure shown in Figure 70 was caused by a series of procedural errors and the use of a home-made pressure gauge, without a stop for its needle, that allowed the aircraft to be over-pressurized on the ground.

Figure 70 A KC-135 aircraft being pressurized at ground level

Interactions between people in complex systems illustrate how problematic comforting terms such as 'condition of incredibility' are, Figure 71.

Figure 71 Who determines the strength of a part?

Implicit in these examples is that there is a natural tension between commercial and engineering safety imperatives: keeping the plant running and the plane flying is the commercial pressure that is resisted by an engineering assessment of risk. This tension between commercial pressures and regulated engineering safety systems is part of everyday life. Beneath the suburban street, high-pressure gas mains run at a lower pressure than those in the countryside because the consequences of failure are likely to be more important in the city. These mains are not just arteries through which to deliver gas; they are a major component of the gas storage system in the UK. If the pressure were to be increased, Transco, the company responsible for gas transmission, would gain an enormous increase in storage capacity at no extra cost. It is the business of Transco to convince the regulator and it is the business of the regulator to be hard to convince.

When the tension breaks down the results are terrible. On privatization the UK rail system transmogrified from a culture of engineering knowledge into a commercial one. This resulted in the Hatfield train crash ushering in a series of delays and spending that put the rail operator, Railtrack, into insolvency. In November 2000 Gerald Corbett, the then chief executive of Railtrack, explained in an interview:

> *The way the maintenance was contracted out broke the engineering chain and, at that time, no one realised that. Under British Rail, the engineering function was very strong. There was a chain going right down to the local area engineer. The local engineer was in charge of the system and optimising the wheel–rail interface and the chain of command went up to the top engineer. But when maintenance was broken off, a load of engineers went into the maintenance companies and some stayed with Railtrack, but the interface stopped being an engineering interface. It became a commercial interface, with all the problems.*
>
> *(Wolmar, 2001)*

More stable industries are aware of the problems of risk and, perhaps more importantly, the public perception of risk. Aircraft accidents are rare but at least the industry has some statistics from which to calculate risk, unlike the nuclear industry which has to calculate risk without much basic data from accidents. However, air travel is becoming more common and so the public perception of risk might increase, as news of crashes becomes more common, even without an increase in the rate of accidents.

An increasingly litigious public demands someone to blame in a risk-free environment; 'The future is a foreign country, they do things differently there' to quote L. P. Hartley (1953).

References

Bacon. F. (1626) *Sylva Sylvarum, or, a Naturall History in Ten Centuries*, printed in London by J. Haviland for William Lee.

British Energy Generation Ltd. (2000) *Assessment of the Integrity of Structures Containing Defects, R6-Revision 4.*

BS 7910 (1999) *Guide on methods for assessing the acceptability of flaws in metallic structures*, British Standards Institute.

CEGB (1976) *Assessment of the Integrity of Structures Containing Defects, CEGB Report R/H/R6*, Revision 1, 1977; Revision 2, 1980; Revision 3, 1986.

Charpy, G. (1909). Report on Impact Tests of Metals. Vth Congress, Copenhagen, International Association for Testing Materials.

Chell, G. G. (1979) 'Elastic–Plastic Fracture Mechanics', in *Developments in Fracture Mechanics-1*, Applied Science Publishers.

Donne, J. (1644) Biathanatos, printed by J. Dawson, London.

Dugdale, D. S. (1960) 'Yielding of steel sheets containing slits', *Journal of the Mechanics and Physics of Solids*, Vol. 8, pp.100–108.

Hall, J. and Goranson, U. G. (1982) 'Principles of Achieving Damage Tolerance with Flexible Maintenance Programs for New and Aging Aircraft', presented at 13th Congress of the International Council of the Aeronautical Sciences/AIAA Aircraft Systems and Technology Conferences, Seattle, Washington, USA, August 26.

Harrison, R. P., K. Loosemoore, et al. (1980) *Assessment of the Integrity of Structures Containing Defects, R6-Revision 2*, CEGB.

Hartley, L. P. (1953) *The Go-Between*, H. Hamilton.

Holland, P. (trans) (1601) *The Historie of the World. Commonly Called The Natural Historie of C. Plinius Secundus*, printed by Adam Islip, London.

Henry, W. (1826) *The elements of experimental chemistry* (10ed), Printed for Baldwin, Cradock and Joy.

Hutchinson, J. W. (1968) 'Singular behaviour at the end of a tensile crack in a hardening material', *Journal of the Mechanics and Physics of Solids*, Vol. 16, pp.13–31.

Hutton, E. (1996) *Sir Henry Bessemer, F. R. S.: An Autobiography*, available online at http://www.history.rochester.edu/ehp-book/shb/ [accessed 11 July 2003].

Irwin, G. R. and Kies, J. A. (1954) 'Critical energy rate analysis of fracture strength', *Welding Journal Research Supplement*, 33(4).

Kirkaldy, D. (1862) Results of an experimental inquiry into the comparative tensile strength and other properties of various kinds of wrought-iron and steel, Bell & Bain.

Kumar V., German M. D. and Shih C. F. (1981) *An engineering approach for elastic–plastic fracture analysis.* EPRI Report NP-1931.

Laham, S. A. (1999) *Stress Intensity Factor and Limit Load Handbook*, British Energy Generation Ltd.

Le Chatelier, H. L. (1904). Contribution à l'étude de la fragilité, Société d'Encouragement pour l'Industrie National.

National Academy of Sciences (1996) 'Egon Orowan' by F. R. N. Nabarro and A. S. Argon in *Biographical Memoirs*, Vol. 70.

Orowan, E. (1949) 'Fracture and strength of solids', *Reports on Progress in Physics*, Vol. 12, pp. 185–232.

Paris, P. C., Gomez, M. P. and Anderson, W. E. (1961) 'A rational analytic theory of fatigue', *The Trend in Engineering at the University of Washington*, Vol. 13, No. 1, pp. 9–14.

Rice, J. R. and Rosengren, G. F. (1968) 'Plane strain deformation near a crack tip in a power law hardening material, *Journal of the Mechanics and Physics of Solids*, Vol. 16, pp.1–12.

Rossmanith, H. P., ed. (1997) *Fracture Research in Retrospect: An Anniversary Volume in Honour of G. R. Irwin's 90th Birthday*, Balkema.

Webster, S. E. and Bannister, A. C. (1999) *Final Publishable Report*, European Seminar on Structural Integrity Assessment Procedures for European Industry (SINTAP). S. Consortium, GKSS Research Center.

Wöhler, A. (1871) *Uber die Festigkeitsversuche mit Eisen und Stahl*, Berlin.

Wolmar, C. (2001) *Broken Rails*, Aurum Press.

Acknowledgements

Photographs and illustrations

Grateful acknowledgement is made to the following sources for permission to reproduce material within this book:

Figure 5: Courtesy of The Principal and The Fellows of Newnham College, Cambridge.

Figure 3: © Science Photo Library.

Figure 4: Courtesy of Watertown Free Public Library.

Figure 6 and cover: Rex Features/Robert Nichols.

Figure 7: Courtesy of Sheffield Forgemasters.

Figure 8: © The British Museum.

Figure 16: La Bibliothèque, Ecole des Mines de Paris.

Figure 18: Courtesy of Professor M. Neil James, Plymouth University and Alan Stone, Aston Metallurgical Services.

Figures 19, 20, 26–29, 34, 35, 42, and 45—48: Courtesy of Adrian Demaid.

Figure 22: Hancock, J. W. and Cowling, M. J., *Metal Science,* Vol. 14, 1980.

Figures 23 and 24: Courtesy of TWI Limited.

Figures 36 and 37: Courtesy of Naval Research Laboratory, USA.

Figure 40: Courtesy of Paul C. Paris.

Figure 43: © Construction Photography.

Figure 60: Armand Dionne/Harvard University/Courtesy of Jim Rice.

Every effort has been made to contact copyright owners. If any have been inadvertently overlooked, the publishers will be happy to make the necessary arrangements at the first opportunity.

Course team acknowledgements

This book was written by Adrian Demaid, with contributions from Michael Fitzpatrick, Nicholas Braithwaite and Lara Mynors, all members of the course team for the Open University course T207 *Engineering: mechanics, materials, design.*

Adrain Demaid thanks Brian Gearing, Martin Goldthorpe, Ian Howard, William Lennox, Ian Milne, Paul Paris, Jim Rice, Alan Rosenfield, Simon Walker and John Yates for their support.

Index